Residue Number Systems
Theory and Implementation

Advances in Computer Science and Engineering: Texts

Editor-in-Chief: Erol Gelenbe *(Imperial College)*
Advisory Editors: Manfred Broy *(Technische Universitaet Muenchen)*
Gérard Huet *(INRIA)*

Published

Vol. 1 Computer System Performance Modeling in Perspective:
A Tribute to the Work of Professor Kenneth C. Sevcik
edited by E. Gelenbe (Imperial College London, UK)

Vol. 2 Residue Number Systems: Theory and Implementation
*by A. Omondi (Yonsei University, South Korea) and
B. Premkumar (Nanyang Technological University, Singapore)*

Advances in Computer Science and Engineering: Texts   Vol. 2

RESIDUE NUMBER SYSTEMS
Theory and Implementation

Amos Omondi
School of Electrical and Electronic Engineering
Yonsei University, South Korea

Benjamin Premkumar
School of Computer Engineering
Nanyang Technological University, Singapore

Imperial College Press

Published by

Imperial College Press
57 Shelton Street
Covent Garden
London WC2H 9HE

Distributed by

World Scientific Publishing Co. Pte. Ltd.
5 Toh Tuck Link, Singapore 596224
USA office: 27 Warren Street, Suite 401-402, Hackensack, NJ 07601
UK office: 57 Shelton Street, Covent Garden, London WC2H 9HE

British Library Cataloguing-in-Publication Data
A catalogue record for this book is available from the British Library.

Advances in Computer Science and Engineering: Texts – Vol. 2
RESIDUE NUMBER SYSTEMS
Theory and Implementation

Copyright © 2007 by Imperial College Press

All rights reserved. This book, or parts thereof, may not be reproduced in any form or by any means, electronic or mechanical, including photocopying, recording or any information storage and retrieval system now known or to be invented, without written permission from the Publisher.

For photocopying of material in this volume, please pay a copying fee through the Copyright Clearance Center, Inc., 222 Rosewood Drive, Danvers, MA 01923, USA. In this case permission to photocopy is not required from the publisher.

ISBN-13 978-1-86094-866-4
ISBN-10 1-86094-866-9

Printed in Singapore.

To my parents-in-law,
Ellen and William Hayes,
for their unflagging support during some trying times.
(*Amos Omondi*)

To
my wife **Manchala** and our sons **Josh** and **Brian**
(*Benjamin Premkumar*)

Preface

Residue number systems (RNS) were invented by the third-century Chinese scholar Sun Tzu—a different Sun Tzu from the author of the famous *Art of War*. In Problem 26 of his *Suan Ching* (*Calculation Classic*), our Sun Tzu posed a mathematical riddle:

```
We have things of which we do not know the number
If we count them by threes, we have two left over
If we count them by fives, we have three left over
If we count them by sevens, we have two left over
How many things are there?
```

In other words, "What number yields the remainders 2, 3, and 2 when divided by 3, 5, and 7, respectively?." In modern terminology, 2, 3, and 2 are *residues*, and 3, 5, and 7, are *moduli*. Sun Tzu gave a rule, the *Tai-Yen* (*Great Generalization*) for the solution of his puzzle. In 1247, another Chinese mathematician, Qin Jiushao, generalized the Great Generalization into what we now call the *Chinese Remainder Theorem*, a mathematical jewel.

In the 1950s, RNS were rediscovered by computer scientists, who sought to put them to use in the implementation of fast arithmetic and fault-tolerant computing. Three properties of RNS make them well suited for these. The first is absence of carry-propagation in addition and multiplication, carry-propagation being the most significant speed-limiting factor in these operations. The second is that because the residue representations carry no weight-information, an error in any digit-position in a given representation does not affect other digit-positions. And the third is that there is no significance-ordering of digits in an RNS representation, which means that faulty digit-positions may be discarded with no effect other than a reduction in dynamic range.

The new interest in RNS was not long-lived, for three main reasons: One, a complete arithmetic unit should be capable of at least addition, multiplication, division, square-root, and comparisons, but implementing the last three in RNS is not easy; two, computer technology became more reliable; and, three, converting from RNS notation to conventional notation, for "human consumption", is difficult. Nevertheless, in recent years there has been renewed interest in RNS. There are several reasons for this new interest, including the following. A great deal of computing now takes place in embedded processors, such as those found in mobile devices, and for these high speed and low-power consumption are critical; the absence of carry-propagation facilitates the realization of high-speed, low-power arithmetic. Also, computer chips are now getting to be so dense that full testing will no longer be possible; so fault-tolerance and the general area of computational integrity have again become more important. Lastly, there has been progress in the implementation of the difficult arithmetic operations. True, that progress has not been of an order that would justify a deluge of letters home; but progress is progress, and the proper attitude should be gratitude for whatever we can get. In any case, RNS is extremely good for many applications—such as digital signal processing, communications engineering, computer security (cryptography), image processing, speech processing, and transforms—in which the critical arithmetic operations are addition and multiplication.

This book is targeted at advanced university students, researchers, and professional engineers interested in RNS and their applications. Both theoretical and practical aspects of RNS are discussed. Other than a basic knowledge of digital logic and some of that fuzzy quality known as "mathematical maturity", no other background is assumed of the reader; whatever is required of conventional arithmetic has been included. It has not been our intention to give the last word on RNS—he or she who must know everything will find satisfaction in the many published papers that are referred to in the book—but, taken as a summary, the book takes the reader all the way from square-one right to the state-of-the-art.

Chapter 1 is an introduction to the basic concepts of RNS, arithmetic in RNS, and applications of RNS. Other conventional and unconventional number systems are also discussed, as, ultimately, it is these that form the basis of implementations of residue arithmetic.

Chapter 2 covers the mathematical fundamentals on which RNS are founded. The main subjects are the congruence relationship (which relates numbers in modular arithmetic), the basic representation of numbers, an algebra of residues, the Chinese Remainder Theorem (very briefly), complex-number representation, and, also very briefly, the detection and correction of errors. The *Core Function*, a useful tool in dealing with the problematic (i.e. hard-to-implement) operations in residue number systems, is also discussed.

Forward conversion, the process of converting from conventional representations to RNS ones, is the main subject of Chapter 3. Algorithms and hardware architectures are given. The chapter is divided into two parts: one for arbitrary moduli-sets and one for "special" moduli-sets—those of the form $\{2^n - 1, 2^n, 2^n + 1\}$—and extensions of these. The latter moduli-sets are of interest because with them implementations of almost all operations are relatively quite straightforward.

Chapters 4, 5, and 6 deal with addition/subtraction, multiplication, and division respectively. (Chapter 6 also includes a discussion of other operations that are closely related to division.) Each chapter has two main parts: one on the conventional version of the operation and one that shows how the conventional algorithms and hardware architectures can be modified to implement the RNS operation. The RNS part is again divided into one part for arbitrary moduli-sets and one part for special moduli-sets.

Reverse conversion, the process of converting back from RNS representations into conventional notations, is covered in Chapter 7. (The general structure of that chapter is similar to that of Chapter 3.) The main topics are the Chinese Remainder Theorem, *Mixed-Radix Conversion* (in which one views a residue number system as corresponding to a conventional system with multiple bases), and the Core Function.

The last chapter is a very brief introduction to some applications of RNS—digital signal processing, fault-tolerant computing, and communications. Lack of space has prevented us from giving an in-depth treatment of these topics. As there are more things in heaven and earth than are dreamt of in our philosophy, we leave it to the imaginative reader to dream up some more applications.

Acknowledgements

In all things, we thank and praise The Almighty. The love and joy I get from my family—Anne, Miles, and Micah—sustain me always. My contribution to this book has been made during my employment at Yonsei Unversity, in a position funded by the Korean government's Institute for Information Technology Assessment. I am indebted to both organizations.

Amos Omondi

I gratefully acknowledge the support rendered by Dr. Omondi, during the course of writing this book for his many insightful comments and suggestions. I also owe my gratitude to my colleagues at the School of Computer Engineering, Nanyang Technological University, Singapore. I am greatly indebted to my wife Manchala, our sons Josh and Brian and to my mother and my family for their love, endurance and patience. Finally, I would like thank my Lord Jesus Christ for His enduring love.

Benjamin Premkumar

Contents

Preface   vii

1. Introduction   1
   - 1.1 Conventional number systems 2
   - 1.2 Redundant signed-digit number systems 5
   - 1.3 Residue number systems and arithmetic 6
     - 1.3.1 Choice of moduli 9
     - 1.3.2 Negative numbers 10
     - 1.3.3 Basic arithmetic 11
     - 1.3.4 Conversion 13
     - 1.3.5 Base extension 14
     - 1.3.6 Alternative encodings 14
   - 1.4 Using residue number systems 15
   - 1.5 Summary 17
   - References 18

2. Mathematical fundamentals   21
   - 2.1 Properties of congruences 22
   - 2.2 Basic number representation 24
   - 2.3 Algebra of residues 27
   - 2.4 Chinese Remainder Theorem 39
   - 2.5 Complex residue-number systems 40
   - 2.6 Redundant residue number systems 42
   - 2.7 The Core Function 44
   - 2.8 Summary 47
   - References 47

3. Forward conversion 49
   3.1 Special moduli-sets 50
      3.1.1 $\{2^{n-1}, 2^n, 2^{n+1}\}$ moduli-sets 52
      3.1.2 Extended special moduli-sets 56
   3.2 Arbitrary moduli-sets: look-up tables 58
      3.2.1 Serial/sequential conversion 59
      3.2.2 Sequential/parallel conversion: arbitrary partitioning 62
      3.2.3 Sequential/parallel conversion: periodic partitioning 65
   3.3 Arbitrary moduli-sets: combinational logic 68
      3.3.1 Modular exponentiation 68
      3.3.2 Modular exponentiation with periodicity 78
   3.4 Summary 80
   References 80

4. Addition 83
   4.1 Conventional adders 84
      4.1.1 Ripple adder 85
      4.1.2 Carry-skip adder 88
      4.1.3 Carry-lookahead adders 91
      4.1.4 Conditional-sum adder 97
      4.1.5 Parallel-prefix adders 101
      4.1.6 Carry-select adder 108
   4.2 Residue addition: arbitrary modulus 111
   4.3 Addition modulo $2^n - 1$ 119
      4.3.1 Ripple adder 122
      4.3.2 Carry-lookahead adder 123
      4.3.3 Parallel-prefix adder 127
   4.4 Addition modulo $2^n + 1$ 130
      4.4.1 Diminished-one addition 130
      4.4.2 Direct addition 131
   4.5 Summary 134
   References 134

5. Multiplication 137
   5.1 Conventional multiplication 138
      5.1.1 Basic binary multiplication 139
      5.1.2 High-radix multiplication 142

5.2	Conventional division	151
	5.2.1 Subtractive division	151
	5.2.2 Multiplicative division	160
5.3	Modular multiplication: arbitrary modulus	162
	5.3.1 Table lookup	162
	5.3.2 Modular reduction of partial products	165
	5.3.3 Product partitioning	169
	5.3.4 Multiplication by reciprocal of modulus	173
	5.3.5 Subtractive division	176
5.4	Modular multiplication: modulus $2^n - 1$	177
5.5	Modular multiplication: modulus $2^n + 1$	185
5.6	Summary	191
	References	191

6. Comparison, overflow-detection, sign-determination, scaling, and division   193

6.1	Comparison	194
	6.1.1 Sum-of-quotients technique	195
	6.1.2 Core Function and parity	197
6.2	Scaling	198
6.3	Division	201
	6.3.1 Subtractive division	201
	6.3.2 Multiplicative division	207
6.4	Summary	210
	References	210

7. Reverse conversion   213

7.1	Chinese Remainder Theorem	213
	7.1.1 Pseudo-SRT implementation	220
	7.1.2 Base-extension implementation	223
7.2	Mixed-radix number systems and conversion	227
7.3	The Core Function	234
7.4	Reverse converters for $\{2n-1, 2n, 2n+1\}$ moduli-sets	237
7.5	High-radix conversion	248
7.6	Summary	251
	References	251

8. Applications   255

8.1	Digital signal processing		256
	8.1.1	Digital filters	257
	8.1.2	Sum-of-products evaluation	264
	8.1.3	Discrete Fourier Transform	272
	8.1.4	RNS implementation of the DFT	275
8.2	Fault-tolerance		278
8.3	Communications		286
8.4	Summary		288
References			289

Index   293

Chapter 1

Introduction

Our main aim in this chapter is to introduce the basic concepts underlying residue number systems (RNS), arithmetic, and applications and to give a brief but almost complete summary. We shall, however, start with a discussion of certain aspects of more commonly used number systems and then review the main number systems used in conventional computer arithmetic. We shall also briefly discuss one other unconventional number system that has found some practical use in computer arithmetic; this is the *redundant*[1] *signed-digit* number system. We have two objectives in these preliminary discussions. The first is to facilitate a contrast between RNS and commonly used number systems. The second is to to recall a few basic properties of the conventional number systems, as, ultimately, it is these that form the basis of implementations of residue arithmetic. The subsequent introduction to RNS consists of some basic definitions, a discussion of certain desirable features of a residue number system, and a discussion of the basic arithmetic operations.

A basic number system consists of a correspondence between sequences of digits and numbers. In a *fixed-point* number system, each sequence corresponds to exactly one number[2], and the *radix-point* —the "decimal point" in the ordinary decimal number system— that is used to separate the integral and fractional parts of a representation is in a fixed position. In contrast, in a *floating-point* number system, a given sequence may correspond to several numbers: the position of the radix-point is not fixed, and each position in a digit-sequence indicates the particular number represented. Usually, floating-point systems are used for the representation

---

[1] In a redundant number system, a given number may have more than one representation.
[2] Notable exceptions occur in certain unconventional number systems.

of real numbers, and fixed-point systems are used to represent integers (in which the radix point is implicitly assumed to be at the right-hand end) or as parts of floating-point representations; but there are a few exceptions to this general rule. Almost all applications of RNS are as fixed-point number systems.

If we consider a number[3] such as 271.834 in the ordinary decimal number system, we can observe that each digit has a *weight* that corresponds to its *position*: hundred for the 2, ten for the 7, ..., thousandth for the 4. This number system is therefore an example of a *positional* (or *weighted*) number system; residue number systems, on the other hand, are non-positional. The decimal number system is also a *single-radix* (or *fixed-radix*) system, as it has only one base (i.e. ten). Although *mixed-radix* (i.e. multiple-radix) systems are relatively rare, there are a few useful ones. Indeed, for the purposes of conversion to and from other number systems, as well as for certain operations, it is sometimes useful to associate a residue number system with a weighted, mixed-radix number system.

1.1 Conventional number systems

In this section we shall review the three standard notations used for fixed-point computer arithmetic and then later point out certain relationships with residue arithmetic.

In general, numbers may be signed, and for binary digital arithmetic there are three standard notations that have been traditionally used for the binary representation of signed numbers. These are *sign-and-magnitude*, *one's complement*, and *two's complement*. Of these three, the last is the most popular, because of the relative ease and speed with which the basic arithmetic operations can be implemented. Sign-and-magnitude notation has the convenience of having a sign-representation that is similar to that used in ordinary decimal arithmetic. And one's complement, although a notation in its own right, more often appears only as an intermediate step in arithmetic involving the other two notations.

The sign-and-magnitude notation is derived from the conventional written notation of representing a negative number by prepending a sign to a magnitude that represents a positive number. For binary computer hardware, a single bit suffices for the sign: a sign bit of 0 indicates a positive

---

[3]For clarity of expression, we shall not always distinguish between a number and its representation.

number, and a sign bit of 1 indicates a negative number. For example, the representation of the number positive-five in six bits is 000101, and the corresponding representation of negative-five is 100101. Note that the representation of the sign is independent of that of the magnitude and takes up exactly one bit; this is not the case both with one's complement and two's complement notations.

Sign-and-magnitude notation has two representations, 000...0 and 100...0, for the number zero; it is therefore redundant. With one exception (in the context of floating-point numbers) this existence of two representations for zero can be a nuisance in an implementation. Addition and subtraction are harder to implement in this notation than in one's complement and two's complement notations; and as these are the most common arithmetic operations, true sign-and-magnitude arithmetic is very rarely implemented.[4]

In one's complement notation, the representation of the negation of a number is obtained by inverting the bits in its binary representation; that is, the 0s are changed to 1s and the 1s are changed to 0s. For example, the representation of the number positive-five in six bits is 000101 and negative-five therefore has the representation 111010. The leading bit again indicates the sign of the number, being 0 for a positive number and 1 for a negative number. We shall therefore refer to the most significant digit as the *sign bit*, although here the sign of a negative number is in fact represented by an infinite string of 1s that in practice is truncated according to the number of bits used in the representations and the magnitude of the number represented. It is straightforward to show that the n-bit representation of the negation of a number N is also, when interpreted as the representation of an unsigned number, that of $2^n - 1 - N$. (This point will be useful in subsequent discussions of basic residue arithmetic.) The one's complement system too has two representations for zero—00...0 and 11...1—which can be a nuisance in implementations. We shall see that a similar problem occurs with certain residue number systems. Addition and subtraction in this notation are harder to implement than in two's complement notation (but easier than in sign-and-magnitude notation) and multiplication and division are only slightly less so. For this reason, two's complement is the preferred notation for *implementing* most computer arithmetic.

---

[4]It is useful to note here that the notation of representation and the notation for the actual arithmetic implementation need not be the same.

Negation in two's complement notation consists of a bit-inversion (that is, a translation into the one's complement) followed by the addition of a 1, with any carry from the addition being ignored. Thus, for example, the result of negating 000101 is 111011. As with one's complement notation, the leftmost bit here too indicates the sign: it is 0 for a positive number and 1 for a negative number; but again, strictly, the sign is actually represented by the truncation of an infinite string. For n-bit representations, representing the negation of the number N may also be viewed as the representation of the positive number $2^n - N$.

In contrast with the first two conventional notations, the two's complement has only one representation for zero, i.e. 00...0. The two's complement notation is the most widely used of the three systems, as the algorithms and hardware designs required for its implementation are quite straightforward. Addition, subtraction, and multiplication are relatively easy to implement with this notation, and division is only slightly less so.

All of the notations above can be readily extended to non-binary radices. The extension of binary sign-and-magnitude to an arbitrary radix, r, involves representing the magnitude in radix-r and using 0 in the sign digit for positive numbers and $r - 1$ for negative numbers. An alternative representation for the sign is to use half of the permissible values of the sign digit (that is, $0 \ldots r/2 - 1$, assuming r is even) for the positive numbers and the other half (that is, $r/2 \ldots r - 1$, for an even radix) for the negative numbers. The generalization of one's complement to an arbitrary radix is known as *diminished-radix complement*, the name being derived from the fact that to negate a number in this notation, each digit is subtracted from the radix diminished by one, i.e. from $r - 1$. Alternatively, the representation of the negation may also be viewed as the result of subtracting the number from $r^n - 1$, where n is the number of digits used in the representations. Thus, for example, the negation of 01432 in radix-8 is 76345, i.e. 77777 − 01432. The sign digit will be 0 for a positive number and $r - 1$ for a negative number. The generalization of two's complement to an arbitrary radix is known as *radix complement notation*. In radix complement notation, the radix-r negation of a number is obtained, essentially, by subtracting from r^n, where n is the number of digits used in the representations. Alternatively, negation may also be taken as the formation of the diminished-radix complement followed by the addition of a 1. Thus, for example, the radix-8 negation of 01432 is 76346, i.e. 100000 − 01432 or 77777 − 01432 + 1. The determination of sign is similar to that for the radix-r diminished-radix complement.

Residue number systems are more complex than the three standard notations reviewed above, and, consequently, cannot be implemented directly with the same efficiency as the conventional arithmetic. Therefore, in practice, residue arithmetic is often realized in terms of lookup-tables (to avoid the complex combinational-logic circuits) and conventional arithmetic (i.e. arithmetic in some standard notation). For example, the sign-and-magnitude approach may be convenient for representing signed numbers in RNS, but actual arithmetic operations might be best realized in terms of radix-complement (two's complement) arithmetic. We shall also see that certain choices of representational parameters in RNS naturally lead to diminished-radix complement (one's complement) arithmetic.

1.2 Redundant signed-digit number systems

Many unconventional number systems have been proposed, and some even have been put to practical use. Nevertheless, very few have had widespread or sustained application. Therefore, other than RNS, we shall restrict ourselves here to just the *redundant signed-digit* (RSD) systems, which have long been used for high-speed arithmetic. As far as practical unconventional notations go, RSD systems are the only serious competitor to RNS. RSD systems may also have some applications in the implementation of residue arithmetic, just as the conventional notations do; this is a largely unexplored area, although there has been some relevant work.

Two of the conventional systems discussed above have some *redundancy*, which means that in each there is at least one number with two or more representations; but the redundancy there is hardly of a useful variety. In contrast, redundant signed-digit number systems have much greater degrees of deliberate and useful redundancy. These systems are mainly used for high-speed arithmetic, especially in multiplication (in which they are used indirectly) and division (in which they are used more directly). They are also used in some algorithms for the evaluation of elementary (i.e. transcendental) functions.

In a typical RSD system, the number of values that any one digit may assume exceeds the value of the radix, and the digits are individually signed. If such a system employs a radix r, then the number of distinct values that a digit may assume lies between $r+1$ and $2r-1$. (In conventional number systems, such as those above, the size of the digit-set does not exceed r.) The digit-set in a redundant signed-digit number system is

$\{-a, -(a-1), \ldots, -1, 0, 1, \ldots, b-1, b\}$, where $\lceil (r-1)/2 \rceil \leq a, b \leq r-1$; usually $a = b$. In contrast, for a conventional number system the digit-set is usually $\{0, 1, \ldots, r-1\}$.

Given the above, it is evident that in an RSD system a given number will have several distinct representations. For example, in the radix-2 RSD system, which employs the digit-set $\{\bar{1}, 0, 1\}$ (where $\bar{1}$ stands for -1), three five-bit representations of the number eleven are 01011 (that is, 8+2+1), $10\bar{1}0\bar{1}$ (that is, 16−4+1), and $0110\bar{1}$ (that is, 8+4−1). The representation of the negation of a number represented in a redundant signed-digit notation is obtained by changing the sign of every non-zero digit. For example, three representations, obtained from the preceding example, of negative-eleven are $0\bar{1}0\bar{1}\bar{1}$ (that is, $-8 - 2 - 1$), $\bar{1}0101$ (that is, $-16 + 4 + 1$), and $0\bar{1}\bar{1}01$ (that is, $-8 - 4 + 1$).

In high-speed multiplication, RSD systems appear implicitly in multiplier recoding [7]. For example, straightforward multiplication by the positive number 11111111 requires eight additions, but only two are required if the multiplier is recoded (on-the-fly) into $\bar{1}00000001$. For division, generating a quotient in RSD notation helps speed up the process by allowing the individual quotient digits to be formed from only approximate comparisons of the partial dividend and the divisor. The redundancy permits later correction of any errors in the choice of digits. For example, if two successive quotient-digits should be 01, but the first digit is guessed to be 1, a correction can subsequently be made by selecting the next digit as $\bar{1}$, since 01 and $1\bar{1}$ represent the same number. The RSDs also find some use in the evaluation of elementary functions, in a manner similar to that of their use in division; the main algorithms here are the CORDIC ones [4]. As far as RNS goes, there has been a few proposals to use RSD to speed up the implementations of RNS-arithmetic [3].

1.3 Residue number systems and arithmetic

Residue number systems are based on the *congruence* relation, which is defined as follows. Two integers a and b are said to be *congruent modulo* m if m divides exactly the difference of a and b; it is common, especially in mathematics tests, to write $a \equiv b \pmod{m}$ to denote this. Thus, for example, $10 \equiv 7 \pmod 3, 10 \equiv 4 \pmod 3, 10 \equiv 1 \pmod 3$, and $10 \equiv -2 \pmod 3$. The number m is a *modulus* or *base*, and we shall assume that its values exclude unity, which produces only trivial congruences.

If q and r are the quotient and remainder, respectively, of the integer division of a by m—that is, $a = q.m + r$—then, by definition, we have $a \equiv r \pmod{m}$. The number r is said to be the *residue* of a with respect to m, and we shall usually denote this by $r = |a|_m$. The set of m smallest values, $\{0, 1, 2, \ldots, m-1\}$, that the residue may assume is called the set of *least positive residues modulo m*. Unless otherwise specified, we shall assume that these are the only residues in use.

Suppose we have a set, $\{m_1, m_2, \ldots, m_N\}$, of N positive and pairwise relatively prime moduli[5]. Let M be the product of the moduli. Then every number $X < M$ has a unique representation in the residue number system, which is the set of residues $\{|X|_{m_i} : 1 \leq i \leq N\}$. A partial proof of this is as follows. Suppose X_1 and X_2 are two different numbers with the same *residue-set*. Then $|X_1|_{m_i} = |X_2|_{m_i}$, and so $|X_1 - X_2|_{m_i} = 0$. Therefore $X_1 - X_2$ is the least common multiple (lcm) of m_i. But if the m_i are relatively prime, then their lcm is M, and it must be that $X_1 - X_2$ is a multiple of M. So it cannot be that $X_1 < M$ and $X_2 < M$. Therefore, the set $\{|X|_{m_i} : 1 \leq i \leq N\}$ is unique and may be taken as the representation of X. We shall write such a representation in the form $\langle x_1, x_2, \ldots, x_N \rangle$, where $x_i = |X|_{m_i}$, and we shall indicate the relationship between X and its residues by writing $X \cong \langle x_1, x_2, \ldots, x_N \rangle$. The number M is called the *dynamic range* of the RNS, because the number of numbers that can be represented is M. For unsigned numbers, that range is $[0, M-1]$.

Representations in a system in which the moduli are not pairwise relatively prime will be not be unique: two or more numbers will have the same representation. As an example, the residues of the integers zero through fifteen relative to the moduli two, three, and five (which are pairwise relatively prime) are given in the left half of Table 1.1. And the residues of the same numbers relative to the moduli two, four, and six (which are not pairwise relatively prime) are given in the right half of the same table. Observe that no sequence of residues is repeated in the first half, whereas there are repetitions in the second.

The preceding discussions (and the example in the left-half of Table 1.1) define what may be considered *standard residue number systems*, and it is with these that we shall primarily be concerned. Nevertheless, there are useful examples of "non-standard" RNS, the most common of which are the *redundant residue number systems*. Such a system is obtained by, essentially, adding extra (redundant) moduli to a standard system. The

---

[5]That is, for every j and k, if $j \neq k$, then m_j and m_k have no common divisor larger than unity.

dynamic range then consists of a "legitimate" range, defined by the non-redundant moduli and an "illegitimate" range; for arithmetic operations, initial operands and results should be within legitimate range. Redundant number systems of this type are especially useful in fault-tolerant computing. The redundant moduli mean that digit-positions with errors may be excluded from computations while still retaining a sufficient part of the dynamic range. Furthermore, both the detection and correction of errors are possible: with k redundant moduli, it is possible to detect up to k errors and to correct up to $\lfloor k/2 \rfloor$ errors. A different form of redundancy can be introduced by extending the size of the digit-set corresponding to a modulus, in a manner similar to RSDs. For a modulus m, the normal digit set is $\{0, 1, 2, \ldots, m - 1\}$; but if instead the digit-set used is $\{0, 1, 2, \ldots, \widetilde{m}\}$, where $\widetilde{m} \geq m$, then some residues will have redundant representations. Redundant residue number systems are discussed in slightly more detail in Chapters 2 and 8.

Table 1.1: Residues for various moduli

N	Relatively prime moduli			Relatively non-prime moduli		
	$m_1 = 2$	$m_2 = 3$	$m_3 = 5$	$m_1 = 2$	$m_2 = 4$	$m_3 = 6$
0	0	0	0	0	0	0
1	1	1	1	1	1	1
2	0	2	2	0	2	2
3	1	0	3	1	3	3
4	0	1	4	0	0	4
5	1	2	0	1	1	5
6	0	0	1	0	2	0
7	1	1	2	1	3	1
8	0	2	3	0	0	2
9	1	0	4	1	1	3
10	0	1	0	0	2	4
11	1	2	1	1	3	5
12	0	0	2	0	0	0
13	1	1	3	1	1	1
14	0	2	4	0	2	2
15	1	0	0	1	3	3

1.3.1 Choice of moduli

Ignoring other, more "practical", issues, the best moduli are probably prime numbers—at least from a purely mathematical perspective. A particularly useful property of such moduli is that of "generation". If a modulus, m, is prime, then there is at least one *primitive root* (or *generator*), $p \leq m - 1$, such that the set $\{|p^i|_m : i = 0, 1, 2, \ldots, m - 2\}$ is the set of all the non-zero residues with respect to m. As an example, if $m = 7$, then we may take $p = 3$, since $\{|3^0|_7 = 1, |3^1|_7 = 3, |3^2|_7 = 2, |3^3|_7 = 6, |3^4|_7 = 4, |3^5|_7 = 5\} = \{1, 2, 3, 4, 5, 6\}$; 5 is also a primitive root. Evidently, for such moduli, multiplication and powering of residues may be carried out in terms of simple operations on indices of the power of the primitive root, in a manner similar to the use of logarithms and anti-logarithms in ordinary multiplication. More on the subject will be found in Chapters 2 and 5.

For computer applications, it is important to have moduli-sets that facilitate both efficient representation and balance, where the latter means that the differences between the moduli should be as small as possible.[6] Take, for example, the choice of 13 and 17 for the moduli, these being adjacent prime numbers; the dynamic range is 221. With a straightforward binary encoding, four bits and five bits, respectively will be required to represent the corresponding residues. In the former case, the representational efficiency is 13/16, and in the latter it is 17/32. If instead we chose 13 and 16, then the representational efficiency would be improved—to 16/16 in the second case— but at the cost of reduction in the range (down to 208). On, the other hand, with the better balanced pair, 15 and 16, we would have both better efficiency and greater range: 15/16 and 16/16 for the former, and 240 for the latter.

It is also useful to have moduli that simplify the implementation of the arithmetic operations. This invariably means that arithmetic on residue digits should not deviate too far from conventional arithmetic, which is just arithmetic modulo a power of two. A common choice of prime modulus that does not complicate arithmetic and which has good representational efficiency is $m_i = 2^i - 1$. Not all pairs of numbers of the form $2^i - 1$ are relatively prime, but it can be shown that that $2^j - 1$ and $2^k - 1$ are relatively prime if and only if j and k are relatively prime. Many moduli sets are based on these choices, but there are other possibilities; for example,

---

[6]Unbalanced moduli-sets lead to uneven architectures, in which the role of the largest moduli, with respect to both cost and performance, is excessively dominant. An example of a moduli-set with good balance is $\{2^n - 1, 2^n, 2^n + 1\}$.

moduli-sets of the form $\{2^n - 1, 2^n, 2^n + 1\}$ are among the most popular in use.

In general, then, there are at least four considerations that should be taken into account in the selection of moduli. First, the selected moduli must provide an adequate range whilst also ensuring that RNS representations are unique. The second is, as indicated above, the efficiency of binary representations; in this regard, a balance between the different moduli in a given moduli-set is also important. The third is that, ideally, the implementations of arithmetic units for RNS should to some extent be compatible with those for conventional arithmetic, especially given the "legacy" that exists for the latter. And the fourth is the size of individual moduli: Although, as we shall see, certain RNS-arithmetic operations do not require carries between digits, which is one of the primary advantages of RNS, this is so only between *digits*. Since a digit is ultimately represented in binary, there will be carries between bits, and therefore it is important to ensure that digits (and, therefore, the moduli) are not too large. Low-precision digits also make it possible to realize cost-effective table-lookup implementations of arithmetic operations. But, on the other hand, if the moduli are small, then a large number of them may be required to ensure a sufficient dynamic range. Of course, ultimately the choices made, and indeed whether RNS is useful or not, depend on the particular applications and technologies at hand.

1.3.2 Negative numbers

Some applications require that it be possible to represent negative numbers as well as positive ones. As with the conventional number systems, any one of the radix complement, diminished-radix complement, or sign-and-magnitude notations may be used in RNS for such representation. The merits and drawbacks of choosing one over the other are similar to those for the conventional notations. In contrast with the conventional notations, however, the determination of sign is much more difficult with the residue notations, as is magnitude-comparison. This is the case even with sign-and-magnitude notation, since determining the sign of the result of an arithmetic operation such as addition or subtraction is not easy—even if the signs of the operands are known. This problem, which is discussed in slightly more detail below and in Chapter 6, imposes many limitations on the application of RNS.

Introduction

The extension of sign-and-magnitude notation to RNS involves the use of a single sign-digit or prepending to each residue in a representation an extra bit or digit for the sign; we shall assume the former. For the complement notations, the range of representable numbers is usually partitioned into two approximately equal parts, such that approximately half of the numbers are positive and the rest are negative. Thus, if the moduli used are $m_1, m_2 \ldots, m_N$, then there are $M \stackrel{\Delta}{=} \prod_{i=1}^{N} m_i$ representable numbers, and every representable number, X, satisfies one of two relations:

$$-\frac{M-1}{2} \le X \le \frac{M-1}{2} \quad \text{if } M \text{ is odd}$$

$$-\frac{M}{2} \le X \le \frac{M}{2} - 1 \quad \text{if } M \text{ is even}$$

Then, for complement notation, if $\langle x_1, x_2, \ldots, x_N \rangle$ is the representation of X, where $x_i = |X|_{m_i}$, then the representation of $-X$ is $\langle \overline{x_1}, \overline{x_2}, \ldots, \overline{x_N} \rangle$, where $\overline{x_i}$ is the m_i's-complement, i.e. the radix complement, or the $(m_i - 1)$'s-complement, i.e. the diminished-radix complement, of x_i. For example, with the moduli-set $\{2, 3, 5, 7\}$, the representation of seventeen is $\langle 1, 2, 2, 3 \rangle$ and the radix-complement representation of its negation is $\langle 1, 1, 3, 4 \rangle$, from $\langle 2-1, 3-2, 5-2, 7-3 \rangle$. The justification for taking the complement (negation) of each residue digit is that $|m_i - x_i|_{m_i} = -x_i$.

If we again take $[0, M-1]$ as the nominal range of an RNS, then it will be seen that in the last example $\langle 1, 1, 3, 4 \rangle$ is also the representation of 193. That is, in splitting the range $[0, M-1]$, we take for positive numbers the subrange $[0, M/2 - 1]$ if M is even, or $[0, (M-1)/2]$ if M is odd, and correspondingly, for the negative numbers we take $[M/2, M-1]$ or $[(M+1)/2, M-1]$. This makes sense since $|M - X|_M = -X$ (in the last example $210 - 17 = 193$) and fits in with the discussion in Section 1.1 as to how we may also view diminished-radix complement and radix-complement representations as the representations of positive numbers.

Unless otherwise stated, we shall generally assume that we are dealing with just the positive numbers.

1.3.3 Basic arithmetic

The standard arithmetic operations of addition/subtraction and multiplication are easily implemented with residue notation, depending on the choice of the moduli, but division is much more difficult. The latter is not surprising, in light of the statement above on the difficulties of sign-determination and magnitude-comparison. Residue addition is carried out by individually adding corresponding digits, relative to the modulus for their position. That is, a carry-out from one digit position is not propagated into the next digit position.

If the given moduli are $m_1, m_2, \ldots, m_N$, $X \cong \langle x_1, x_2, \ldots, x_N \rangle$ and $Y \cong \langle y_1, y_2, \ldots y_N \rangle$, i.e. $x_i = |X|_{m_i}$ and $y_i = |Y|_{m_i}$, then we may define the addition $X + Y = Z$ by

$$X + Y \cong \langle x_1, x_2, \ldots, x_N \rangle + \langle y_1, y_2, \ldots y_N \rangle$$
$$= \langle z_1, z_2, \ldots, z_N \rangle \qquad \text{where } z_i = |x_i + y_i|_{m_i}$$
$$\cong Z$$

As an example, with the moduli-set $\{2, 3, 5, 7\}$, the representation of seventeen is $\langle 1, 2, 2, 3 \rangle$, that of nineteen is $\langle 1, 1, 4, 5 \rangle$, and adding the two residue numbers yields $\langle 0, 0, 1, 1 \rangle$, which is the representation for thirty-six in that system.

Subtraction may be carried out by negating (in whatever is the chosen notation) the subtrahend and adding to the minuend. This is straightforward for numbers in diminished-radix complement or radix complement notation. For numbers represented in residue sign-and-magnitude, a slight modification of the algorithm for conventional sign-and-magnitude is necessary: the sign digit is fanned out to all positions in the residue representation, and addition then proceeds as in the case for unsigned numbers but with a conventional sign-and-magnitude algorithm.

Multiplication too can be performed simply by multiplying corresponding residue digit-pairs, relative to the modulus for their position; that is, multiply digits and ignore or adjust an appropriate part of the result. If the given moduli are $m_1, m_2, \ldots, m_N$, $X \cong \langle x_1, x_2, \ldots, x_N \rangle$ and $Y \cong \langle y_1, y_2, \ldots y_N \rangle$, i.e. $x_i = |X|_{m_i}$ and $y_i = |Y|_{m_i}$, then we may define the multiplication $X \times Y = Z$ by

$$X \times Y \cong \langle x_1, x_2, \ldots, x_N \rangle \times \langle y_1, y_2, \ldots y_N \rangle$$
$$= \langle z_1, z_2, \ldots, z_N \rangle \qquad \text{where } z_i = |x_i \times y_i|_{m_i}$$
$$\cong Z$$

As an example, with the moduli-set $\{2, 3, 5, 7\}$, the representation of seventeen is $\langle 1, 2, 2, 3 \rangle$, that of nineteen is $\langle 1, 1, 4, 5 \rangle$, and that of their product, three hundred and twenty-three, is $\langle 1, 2, 3, 1 \rangle$. As with addition, obtaining the modulus with respect to m_i can be implemented without division, and quite efficiently, if m_i is of a suitable form.

Basic fixed-point division consists, essentially, of a sequence of subtractions, magnitude-comparisons, and selections of the quotient-digits. But comparison in RNS is a difficult operation, because RNS is not positional or weighted. Consider, for example, the fact that with the moduli-set $\{2,3,5,7\}$, the number represented by $\langle 0,0,1,1\rangle$ is almost twice that represented by $\langle 1,1,4,5\rangle$, but this is far from apparent. It should therefore be expected that division will be a difficult operation, and it is. One way in which division could be readily implemented is to convert the operands to a conventional notation, use a conventional division procedure, and then convert the result back into residue notation. The conversions are, however, time consuming, and a direct algorithm should be used if possible. The essence of a good RNS division algorithm will therefore be a relatively efficient method of performing magnitude-comparisons. Such algorithms are discussed in Chapter 6. In a sense, all of them require what are essentially conversions out of the RNS, and, compared with conventional division algorithms, all of them are rather unsatisfactory.

1.3.4 Conversion

The most direct way to convert from a conventional representation to a residue one, a process known as *forward conversion*, is to divide by each of the given moduli and then collect the remainders. This, however, is likely to be a costly operation if the number is represented in an arbitrary radix and the moduli are arbitrary. If, on the other hand, the number is represented in radix-2 (or a radix that is a power of two) and the moduli are of a suitable form (e.g. 2^n-1), then there procedures that can be implemented with more efficiency. The conversion from residue notation to a conventional notation, a process known as *reverse conversion*, is more difficult (conceptually, if not necessarily in the implementation) and so far has been one of the major impediments to the adoption use of RNS. One way in which it can be done is to assign weights to the digits of a residue representation and then produce a "conventional" (i.e positional, weighted) mixed-radix representation from this. This mixed-radix representation can then be converted into whatever conventional form is desired. In practice, the use of a direct conversion procedure for the latter can be avoided by carrying out the arithmetic of the conversion in the notation for the result. Another approach involves the use of the Chinese Remainder Theorem, which is the basis for many algorithms for conversion from residue to conventional notation; this too involves, in essence, the extraction of a mixed-radix representation.

1.3.5 Base extension

We have so far assumed that once the moduli-set has been determined, then all operations are carried out with respect to only that set. That is not always so. A frequently occurring computation is that of *base extension*, which is defined as follows. Given a residue representation $\langle |X|_{m_1}, |X|_{m_2}, \ldots, |X|_{m_N} \rangle$ and an additional set of moduli, $m_{N+1}, m_{N+2}, \ldots, m_{N+K}$, such that $m_1, m_2, \ldots m_N, m_{N+1}, \ldots, m_{N+K}$ are all pairwise relatively prime, we want to compute the residue representation $\langle |X|_{m_1}, |X|_{m_2}, \ldots, |X|_{m_N}, |X|_{m_{N+1}}, \ldots, |X|_{m_{N+K}} \rangle$. Base extension is very useful in dealing with the difficult operations of reverse conversion, division, dynamic-range extension, magnitude-comparison, overflow-detection, and sign-determination. The operation is discussed in more detail in Chapters 2 and 7.

1.3.6 Alternative encodings

The basic encoding for values (residues, etc.) in almost all RNS implementations is conventional binary. That is, if the largest possible value is N, then each value is represented in $\lceil \log_2 N \rceil$ bits, and the weighted sum of the bits in a representation is the value represented. This is the encoding we shall assume throughout the book. Nevertheless, alternative encodings are possible, and one that is perhaps worthy of note is *one-hot encoding* (OHE), which is a special case of *n-of-m* encoding[7]. In OHE, if the largest possible value is N, then each data value is represented in N bits. Such a representation for a value n has 0s in all bit positions, except the nth, which is set to 1. Evidently OHE has less representational efficiency than straightforward binary encoding. The basic idea of OHE is not new, but its application is to RNS is. This is because OHE is not suitable for the representation of large values; but in RNS the values generally tend to be small, and so OHE is more practical.

The nominal advantages of OHE residue numbers are as follows [6]. First, a change in the value of the represented number requires changing at most two bits; this is the smallest change possible and is beneficial for power consumption, since, in current technology, that is related to the number of transitional activities. Second, arithmetic operations are simply implemented by shifting: addition consists of rotating one operand by an amount equal to the value of the other operand, and multiplication

---

[7] In *n*-hot (*n*-of-*m*) encoding, a value is represented by 1s on n out of m lines.

(assuming moduli-sets with primitive roots) can be realized by rotations and index ("log") and reverse-index ("anti-log") operations that are simple permutations of wires. (Multiplication where there is no primitive root can be realized by using barrel shifters.) Other operations that are easily realized are inverse-calculation (the analogues, discussed in Chapter 2, of negation and reciprocation in conventional arithmetic), index computation, and modulus conversion. Of course, whether the nominal advantages can be turned into practical advantages is highly dependent on the technology at hand—for example, on whether, for arithmetic operations, barrel shifters can be realized with better efficiency than in conventional circuits. Nevertheless, initial results appear promising.

All that said, OHE does have two significant limitations. First, the sizes of basic arithmetic circuits, such as adders and multipliers, grow at a rate $O(m^2)$, where m is the modulus, in contrast, with conventional binary circuits, in which the rate of growth can be constrained to $O(m \log m)$. Second, the poor representational efficiency means that the cost of interconnections will be much higher than with conventional circuits—at least, $O(m)$ growth versus $O(\log m)$; this may be critical in current technologies, in which interconnections play an increasingly larger role.

1.4 Using residue number systems

We now give an example that demonstrates a typical application of residue number systems: a multiply-accumulate operation over a sequence of scalars. This is an operation that occurs very frequently in digital-signal-processing applications, one of the areas for which RNS is suitable.

EXAMPLE. Let the moduli-set be $\{m_i\} = \{2, 3, 5, 7\}$. The dynamic range of this moduli-set 210. Suppose we wish to evaluate the sum-of-products $7 \times 3 + 16 \times 5 + 47 \times 2$. The residue-sets are

$$|2|_{m_i} = \{0, 2, 2, 2\}$$
$$|3|_{m_i} = \{1, 0, 3, 3\}$$
$$|5|_{m_i} = \{1, 2, 0, 5\}$$
$$|7|_{m_i} = \{1, 1, 2, 0\}$$
$$|16|_{m_i} = \{0, 1, 1, 2\}$$
$$|47|_{m_i} = \{1, 2, 2, 5\}$$

We proceed by first computing the products by multiplying the corresponding residues:

$$|7 \times 3|_{m_i} = \{|1 \times 1|_2, |1 \times 0|_3, |2 \times 3|_5, |0 \times 3|_7\}$$
$$= \{1, 0, 1, 0\}$$

$$|16 \times 5|_{m_i} = \{|0 \times 1|_2, |1 \times 2|_3, |1 \times 0|_5, |2 \times 5|_7\}$$
$$= \{0, 2, 0, 3\}$$

$$|47 \times 2|_{m_i} = \{|1 \times 0|_2, |2 \times 2|_3, |2 \times 2|_5, |5 \times 2|_7\}$$
$$= \{0, 1, 4, 3\}$$

Now that we have computed the products, the sum of products can be evaluated by adding the corresponding residues:

$$|7 \times 3 + 16 \times 5 + 47 \times 2|_{m_i} = \{|1 + 0 + 0|_2, |0 + 2 + 1|_3, |1 + 0 + 4|_5,$$
$$|0 + 3 + 3|_7\}$$
$$= \{1, 0, 0, 6\}$$

END EXAMPLE

The residue representation $\langle 1, 0, 0, 6 \rangle$ corresponds to the decimal 195, which is correct. Consider now the cost of the multipliers required if RNS was not used. We need to multiply 7 and 3, and this requires a 5-bit multiplier. Similarly, the multipliers required for the products of 16 and 5 and 47 and 2 are be 8-bit and 9-bit, respectively. But with the use of RNS, 6-bit multipliers would suffice for all the multiplications. Thus the use of RNS can result in considerable savings in operational-time, area, and power. Moreover, all multiplications can be done in parallel in order to increase the over all speed. This shows a primary advantage of RNS relative to conventional notations. Although there is a final overhead in conversion between the number systems, it may be considered as a one-off overhead.

Whether RNS is useful or not crucially depends on the application. For appropriate applications, there have been some clear evidence of the advantages. As an example, [5] gives the design of a digital-signal processor whose arithmetic units are fully in RNS. An evaluation of this processor shows better performance at lower cost (chip area) than with a conventional system.

1.5 Summary

The main point of this chapter has been to introduce the essential aspects of residue number systems and to give a brief, but essentially complete, summary of what is to come in the rest of the book. We have also reviewed the three standard number systems and one other unconventional number system used for arithmetic in computers; all of these have some relationships with residue arithmetic, especially in implementation. Major differences between residue arithmetic and standard computer arithmetic are the absence of carry-propagation (in the former), in the two main operations of addition and multiplication, and the relatively low precisions required, which leads to practical table-lookup implementations. In practice, these may make residue arithmetic worthwhile, even though in terms of asymptotic bounds such arithmetic has little advantage over conventional arithmetic. A wealth of information on residue number systems, and their applications, will be found in [1, 2], although these may currently be difficult to obtain.

We have noted above that addition and multiplication are easy to realize in a residue number system but that operations that require the determination of magnitudes (e.g. magnitude-comparison, division, overflow, and sign-determination) are difficult; reverse conversion is also difficult.[8] Together, these suggest that the best applications for residue number systems are those with numerous additions and multiplications, but relatively few conversions or magnitude operations, and in which the results of all arithmetic operations lie within a known range. Such applications are typical of digital signal processing, in which, for example, the computation of inner-products (essentially, sequences of multiply-add operations) is quite common.

Residue number systems are also useful in error detection and correction. This is apparent, given the independence of digits in a residue-number representation: an error in one digit does not corrupt any other digits. In general, the use of redundant moduli, i.e. extra moduli that play no role in determining the dynamic range, facilitates both error detection and correction. But even without redundant moduli, fault-tolerance is possible, since computation can still continue after the isolation of faulty digit-positions, provided that a smaller dynamic range is acceptable. Lastly, RNS can help speed up complex-number arithmetic: for example, the multiplication of

---

[8]The essence of the problem with these operations is that they require interactions between all the digits of an RNS representation.

two complex numbers requires four multiplications and two additions when done conventionally but only two multiplications and two additions with the right sort of RNS. (See Chapter 2.)

Figure 1.1 shows the general structure of an RNS processor.

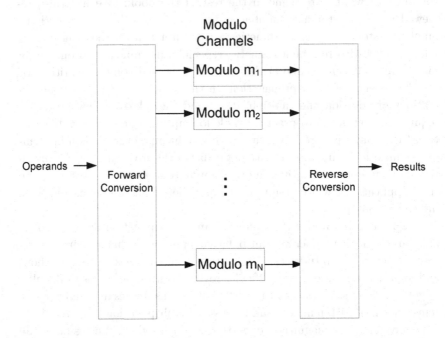

Figure 1.1: General structure of an RNS processor

References

(1) N. S. Szabo and R. I. Tanaka. 1967. *Residue Arithmetic and Its Applications to Computer Technology.* McGraw-Hill, New York.

(2) M. A. Soderstrand et. al. 1986. *Residue Number System Arithmetic: Modern Applications in Digital Signal Processing.* IEEE Press, California.

(3) A. Lindstrom, M. Nordseth, L. Bengtsson, and A. Omondi. 2003. Arithmetic circuits combining residue and signed-digit representations. In: A. R. Omondi and S. Sedhukin, Eds. *Advances in Computer Systems Architecture, Lecture Notes In Computer Science, vol. 2823* (Springer-Verlag, Heidelberg) pp 246–257.

(4) J.-M. Muller. 1997. *Elementary Functions: Algorithms and Implementations.* Birkhauser, Boston.
(5) R. Charles and L. Sousa. 2003. "RDSP: A RISC DSP based on residue number system." In: *Proceedings, Euromicro Symposium on Digital System Design.*
(6) W. A. Chren. 1998. One-hot residue encoding for low delay-power product CMOS design. *IEEE Transactions on Circuits and Systems – II: Analog and Digital Signal Processing*, 45(3):303–313.
(7) A. R. Omondi. 1994. *Computer Arithmetic Systems.* Prentice-Hall, UK.

Chapter 2

Mathematical fundamentals

This chapter consists of several short sections that cover the mathematical fundamentals on which residue number systems are founded; more advanced results are discussed in subsequent chapters, as needed. As we have seen in Chapter 1, the basic relationship between numbers in a residue number system is that of a congruence relative to a given modulus. The first section of the chapter goes into some more detail on the properties of this relation, especially as it pertains to the basic arithmetic operations. The second section discusses the basic representation of numbers and also includes brief discussions of certain characteristics of residue representations. In the third section, a basic algebra of residues is developed. The fourth section briefly introduces one of the most important results in the theory of residue numbers: the Chinese Remainder Theorem. The fifth section covers complex-number representation within residue number systems. The sixth section is an introduction to one of the more promising application-areas for reside number systems: fault detection and correction, through the use of redundant number systems. The Core Function, a useful tool in dealing with the problematic (i.e. hard-to-implement) operations in residue number systems, is discussed in the seventh section. The last section is a concluding summary.

The theorems and properties given of congruences also form an important background in dealing with conversion between number systems; and, as we shall see in subsequent chapters, these results are also important in more practical ways, such as the design of efficient hardware for modular arithmetic operations. Unless otherwise specified, a reference to *residue-set*, with respect to a modulus m, will mean the smallest set of positive residues. Nevertheless it will be evident that most of these properties apply to all residues. The reader need not go through the entire chapter at one sitting but may later return to parts, according to the requirements in subsequent chapters.

2.1 Properties of congruences

In this section, we shall introduce some more fundamental properties of congruences, which properties apply regardless of the numbers and residue-sets at hand. These properties are helpful in understanding residues and the algebra of residues. The basic properties are as follows.

Addition and subtraction

Congruences with respect to the same modulus may be added or subtracted, and the result will be a valid congruence. That is, if

$$X \equiv x \pmod{m}$$
$$Y \equiv y \pmod{m}$$

then

$$X \pm Y \equiv x \pm y \pmod{m}$$

One implication of this property is that terms may be transferred from one side of a congruence to the other by a simple sign-change, just as is the case in the "normal" algebra of numbers.

Multiplication

Congruences with respect to the same modulus may be multiplied, and the result is a valid congruence. That is, if

$$X \equiv x \pmod{m}$$
$$Y \equiv y \pmod{m}$$

then

$$X \times Y \equiv x \times y \pmod{m}$$

It follows from this that raising both sides of the congruence to the same power (a positive integer), or multiplying the congruence by a constant, results in a valid congruence.

Extension of sum and product properties

The properties above for addition and multiplication have two direct extensions. If $\{x_1, x_2, ..., x_N\}$ and $\{y_1, y_2, ..., y_N\}$ are, respectively, the residue-sets (without any restrictions) of X and Y, obtained relative to the moduli $m_1, m_2, \ldots, m_N$, then the residue-set of $X \pm Y$ is

$$\{x_1 \pm y_1, x_2 \pm y_2, \ldots, x_N \pm y_N\}$$

and that for $X \times Y$ is

$$\{x_1 \times y_1, x_2 \times y_2, \ldots \ldots, x_{n-1} \times y_{n-1}\}$$

We can be more rigorous and insist that the sum, or difference, or product lie within the permissible range, that is, $[0, M-1]$, where M is the product of the moduli $m_1, m_2, \ldots, m_N$. We may then rewrite the above extensions as

$$X \pm Y \cong \langle |x_1 \pm y_1|_{m_0}, |x_2 \pm y_2|_{m_1}, \ldots, |x_{n-1} \pm y_N|_{m_N} \rangle$$

and

$$X \times Y \cong \langle |x_1 \times y_1|_{m_0}, |x_2 \times y_2|_{m_1}, \ldots, |x_N \times y_N|_{m_N} \rangle$$

EXAMPLE. Suppose $X = 21$, $Y = 11$ and the moduli-set is $\{2, 3, 5\}$. Then

$$X + Y \cong \langle |x_1 + y_1|_2, |x_2 + y_2|_3, |x_3 + y_3|_5 \rangle$$

where, $x_i = |X|_{m_i}$ and $y_i = |Y|_{m_i}$. And with $X = 21$ and $Y = 11$, we have

$$21 + 11 \cong \langle |1 + 1|_2, |0 + 2|_3, |1 + 1|_5 \rangle$$
$$32 \cong \langle 0, 2, 2 \rangle$$

Similarly, for multiplication, we have

$$X \times Y \cong \langle |x_1 \times y_1|_2, |x_2 \times y_2|_3, |x_3 \times y_3|_5 \rangle$$
$$21 \times 11 \cong \langle |1 \times 1|_2, |0 \times 2|_3, |x_3 \times y_3|_5 \rangle$$

END EXAMPLE

Transitivity

Congruences are transitive. That is, if

$$X \equiv Y \pmod{m}$$
$$Y \equiv Z \pmod{m}$$

then

$$X \equiv Z \pmod{m}$$

Division

Dividing a number, its residue with respect to a given modulus, and the modulus by a non-zero common factor results in a valid congruence. That is, if we have a congruence

$$X \equiv x \pmod{m}$$

then

$$\frac{X}{n} \equiv \frac{x}{n} \pmod{\frac{m}{n}}$$

From this, it follows that dividing a number and its residue by a common factor that is prime relative to the modulus also results in a valid congruence. By the last equation, if

$$aX \equiv ax \pmod{m}$$

where a and m are relatively-prime, then we immediately have another valid congruence:

$$X \equiv x \pmod{m}$$

The above properties show that in residue arithmetic the basic operations, with the exception of division, are distributive over the moduli. This— the potential for parallelism— is one of the aspects that makes residue arithmetic particularly attractive.

2.2 Basic number representation

We now consider the representation of numbers in a residue number system. A key point made here is that for most practical computational purposes, several moduli are required and these moduli should be prime relative to one another. If an insufficient number of moduli is used, or the moduli are not all relatively-prime, then it it is possible to obtain incorrect results because, in the course of computation, it is possible to have two or more numbers that are nominally different but which have the same representation in the residue number system used. These points have already been touched on in Chapter 1, but we shall now consider a detailed example.

Consider a multiply-and-accumulate operation with respect to the modulus 17. All the positive residues will be in the interval $[0, 16]$. Now, take the multiply-and-accumulate operation

$$X = 7 \times 9 + 3 \times 4 + 19 \times 56 + 12 \times 14$$

Reduction, relative to the modulus, of both sides of this equation yields

$$|X|_{17} = |7 \times 9 + 3 \times 4 + 19 \times 56 + 12 \times 14|_{17}$$
$$= |63 + 12 + 1064 + 168|_{17}$$
$$= |1307|_{17}$$
$$= 15$$

Using normal multiplication and addition, we have arrived at the result $X = 1307$, and then computed $|1307|_{17} = 15$. But in a modular multiply-and-accumulate operation, a set of equations such as these are insufficient to uniquely identify the result, even though they do give us some information about X—to wit, that $|X|_{17} = 15$.

In the example just given, there are seventy-six numbers between 0 and 1307 whose residue with respect to 17 is also 15; in other words, there are other sets of operands for which the multiply-and-accumulate gives 15 as the result. In order to narrow down the possibilities, we need to consider evaluating the residue of X with respect to another modulus. Let us take 13 as one such modulus, since this is prime relative to 17. Taking modulus of X with respect to 13 yields 7. The possibilities are now reduced to $\lfloor 76/13 \rfloor = 5$. This, however, is still insufficient because there is still some ambiguity about X. We need to further reduce the possibilities—to either zero or one—and this can be achieved by using additional moduli whose product is 7. Since 7 is prime, we may take that as the additional modulus. We are then in a position to uniquely determine the representation of X in the residue system. Our moduli-set is now $\{7, 13, 17\}$, in which all the elements are pairwise relatively prime. We now compute $|X|_7$, which is the residue 5. So the residue-set of X with respect to this moduli-set is $\{5, 7, 15\}$.

It can be shown that there is one and only one number between 0 and $7 \times 13 \times 17 = 1547$ exists that has the residue-set $\{5, 7, 15\}$. Suppose there was another number, say Y, with the same residue set. Then

$$|X|_{m_i} = x_i \qquad m_i = 7, 13, 17$$
$$|Y|_{m_i} = y_i$$

and subtracting, we have

$$|X - Y|_{m_i} = x_i - y_i$$

Since, by our assumptions (X and Y have the same residue set), $x_i - y_i = 0$, and so

$$|X - Y|_{m_i} = 0$$

If that is so, then $X - Y$ should be divisible each of the m_is. But m_i consists of a set of relatively-prime numbers. Therefore, if the difference, $X - Y$ is divisible by each of the m_is, then that difference must be either 1547 or a multiple of 1547. This shows that in the interval $[0, 1546]$ there cannot be a number such as the postulated Y. Therefore, X is unique.

What we have shown is that in order to uniquely represent any number in a residue number system, we need to have a set of relatively-prime moduli and that the number to be represented should lie in an interval between 0 and the product of the moduli, which product happens to be the least common multiple of the moduli.

We have noted above that residue number systems are particularly advantageous with respect to the arithmetic operations of addition and multiplication, because of the inherent parallelism. Such merits cannot, however, be readily extended to other useful operations, and this limitation has usually constrained the widespread practical application of residue number systems. We have in mind here are operations such as magnitude-comparison, which in a more conventional number system may be taken as a subtraction (essentially an addition) followed by the determination of sign. Consider, for example, the representations of the decimal numbers 34, 67, and 1300 in the RNS with the moduli-set used $\{7, 13, 17\}$. These representations are $\langle 6, 8, 0 \rangle$ for 34, $\langle 4, 2, 16 \rangle$ for 67, and $\langle 5, 0, 8 \rangle$ for 1300. Observe that each of the residues of 1300 is smaller than some residue for 34 or 67, although 1300 is much larger than both 34 and 67. Unlike the situation in a positional, weighted number system, here the positions of the digits give no helpful information. What this example shows is that in the translation to residue representation, all magnitude information is lost. Consequently, it is not surprising that magnitude-comparison with residues numbers is a difficult task. There exist several methods for magnitude-comparison, but most of these are complex and cannot be easily implemented in hardware—at least relative to addition, subtraction, and multiplication. We should therefore expect that division, which in its most basic form consists of comparisons of partial remainders with multiplies of the divisor and the determination of the signs of the partial remainders, will also be similarly problematic. This is indeed the case, and it is so for any operation that explicitly or implicitly requires magnitude-comparison.

In most number systems, the representation of every number should, ideally, be unique. This is evidently the case in a typical positional, weighted number system, such as the ordinary decimal number system. On the other hand, with residue number systems we observe that the residues relative to a given modulus repeat after a definite period, as do the residue-sets relative to a given moduli-set, once the upper limit of the dynamic range has been exceeded. (See Table 1.1 for example.) Thus the number of states that can uniquely be represented is limited in a residue number system. With the moduli-set $\{3, 4, 5\}$, for example, the numbers 47 and 107 both have the same representation, which is $\langle 2, 3, 2 \rangle$. The period of this RNS is $3 \times 4 \times 5 = 60$; so the residues repeat for integer multiples of 60. For example, the residues are same for 47, 107, 167, 227, and so forth.

Given a particular RNS, the direct way to increase the number of permissible states (i.e. the dynamic range) is through the inclusion of more moduli that are pairwise relatively prime to the other members in the given moduli-set and to each other; this essentially yields another RNS. Otherwise, we must restrict ourselves to the basic period (60 in the example above), which corresponds to the dynamic range of the given residue number system.

2.3 Algebra of residues

In this section we introduce several important properties of residue numbers and arithmetic on the same. These properties are foundational for later chapters on arithmetic, conversion and applications. Most of the properties below are immediately obvious follow from the definition of residue numbers; they follow directly from the properties given above of congruences and modular arithmetic.

Additive inverse

The additive inverse, $\bar{x}$, of a residue, x, is defined by the equation

$$x + \bar{x} = 0$$

The additive-inverse operation may be applied to individual residues, or to a system as a whole, and its main role is in subtraction. The additive inverse exists and is unique for any residue. It is obtained through a simple operation:

$$\bar{x} = |m - x|_m$$

(The reader will observe here a direct correspondence with the discussions in Chapter 1 of the representation of negative numbers through complementation.)

EXAMPLE. Consider the moduli-set $\{2,3,5\}$. The residue-set for 24 relative to this set is $\{0,0,4\}$. So the additive inverse is computed as

$$|2-0|_2 = 0$$
$$|3-0|_3 = 0$$
$$|5-4|_5 = 1$$

That is, the additive inverse of $\langle 0,0,4 \rangle$ is $\langle 0,0,1 \rangle$.
END EXAMPLE

Subtraction in an RNS may be defined as the addition of the additive inverse of the subtrahend.

EXAMPLE. Consider the numbers 24 and 35 and the moduli-set $\{2,3,5\}$. The residue-set and that of the corresponding inverse for 35 are $\{1,2,0\}$ and $\{1,1,0\}$ respectively. And those for 24 are $\{0,0,4\}$ and $\{0,0,1\}$ The subtraction $35 - 24$ may therefore be carried out as

$$35 - 24 = \langle 1,2,0 \rangle + \langle \bar{0},\bar{0},\bar{4} \rangle$$
$$= \langle 1,2,0 \rangle + \langle 0,0,1 \rangle$$
$$= \langle 1,2,1 \rangle$$

The set $\{1,2,1\}$ corresponds to 11. And the subtraction $24 - 35$ may be carried out as

$$24 - 35 = \langle 0,0,4 \rangle + \langle \bar{1},\bar{2},\bar{0} \rangle$$
$$= \langle 0,0,4 \rangle + \langle 1,1,0 \rangle$$
$$= \langle 1,1,4 \rangle$$

The residue-set obtained is the additive inverse of the $\langle 1,2,1 \rangle$ and corresponds to -11.
END EXAMPLE

If no additional information is given, then in the last case one could also interpret $\langle 1,1,4 \rangle$ as the representation for 19. This difficulty associated with the RNS can be partially removed by dividing the entire range into two approximately equal subranges: one for X and one for its negation, $\overline{X}$. For example, the dynamic range of the moduli-set $\{2,3,5\}$ is 30, and this may be divided into two subranges, such that the residue representations corresponding to numbers between 0 through 14 are considered positive and those in the range 15 through 29 are considered negative. The ambiguity in interpretation is then eliminated.

Addition and subtraction

$$|X \pm Y|_m = | \ |X|_m \pm |Y|_m \ |_m \qquad (2.1)$$

The proof of this follows from the addition-property of Section 2.1: If $x = |X|_m$ and $y = |Y|_m$, then

$$X + Y \equiv |x + y|_m$$

The congruence relation simply states that $x + y$ is the residue $|X + Y|_m$. According to the definition of residue, $x + y$ should lie within the interval $[0, m - 1]$. Since the sum of individual residues can exceed the modulus, m, a modular reduction is necessary to bring the sum back to within the legitimate range. If y is greater than x, then $x - y$ is negative and can be converted to a positive residue by computing the additive inverse.

EXAMPLE. Take the moduli-set $\{2, 3, 5, 7\}$. Let $X = 27$ and $Y = 145$; the corresponding residue representations are $\langle 1, 0, 2, 6 \rangle$ and $\langle 1, 1, 0, 5 \rangle$ respectively. Whence

$$27 + 145 \cong \langle |1 + 1|_2, |0 + 1|_3, |2 + 0|_5, |6 + 5|_7 \rangle$$
$$= \langle 0, 1, 2, 4 \rangle$$

END EXAMPLE

The addition and subtraction operation in the case of residues is defined as

$$|x \pm y|_m = | \ |x|_m \pm |y|_m \ |_m$$

where x and y are residues of X and Y with respect to modulus m. For example, if $X = 6$ and $Y = 5$, then with respect to the modulus 7, we have

$$|6 + 5|_7 = |11|_7$$
$$= 4$$

Multiplication

$$|X \times Y|_m = | \ |X|_m \times |Y|_m \ |_m \qquad (2.2)$$

This property is similar to that of addition and subtraction. The proof follows from the congruence relation that states that if x and y are the residues of X and Y, with respect to m, then $x \times y$ is the residue of $X \times$

Y (mod m). The product of the individual residues may be greater than m, hence the final (outer) modular reduction.

EXAMPLE. Take $X = 16$ and $Y = 9$. Relative to the moduli-set $\{2, 3, 5, 7\}$, the residue representations of X and Y are $\langle 0, 1, 1, 2 \rangle$ and $\langle 1, 0, 4, 2 \rangle$, respectively. So
$$|16 \times 9|_7 = \langle |0 \times 1|_2, |1 \times 0|_3 |1 \times 4|_5 |2 \times 2|_7 \rangle$$
$$= \langle 0, 0, 4, 4 \rangle$$
END EXAMPLE

It is necessary to insure that the product is within the range of the residue system. Therefore, for the multiplication of residues we have
$$|x \times y|_m = | |x|_m \times |y|_m |_m$$

Multiples of a modulus

$$|k \times m|_m = 0 \qquad \text{for any integer } k$$

The definition of a congruence states that X is congruent to x modulo m if and if only m divides exactly the difference between $X - x$ or x is the residue of X with respect to m. In this case, $k \times m - 0$ is exactly divisible by m. So $|k \times m|_m = 0$ is true, and from the last equation we have an immediate corollary:
$$|kx|_{km} = k|x|_m$$

Addition and subtraction of integer multiple of a modulus

The residue of a sum or a difference of a number and an integer multiple of the modulus is the same as the residue of the number:
$$|X \pm k \times m|_m = |X|_m$$
The proof of this follows from the multiplication property.

EXAMPLE. Take $X = 17$ and the moduli-set $\{2, 3, 5\}$. Then
$$|17 \pm (k \times 5)|_{m_i} = \{|17|_2, |17|_3 |17|_5\}$$
$$= \{1, 2, 2\}$$
where k is an integer.
END EXAMPLE

Multiplicative inverses

The *multiplicative inverse* is an analogue of the reciprocal in conventional arithmetic and is defined as follows.

DEFINITION. If x is a non-zero integer, then x^{-1} is the multiplicative inverse of x, with respect to the modulus m if

$$\left| x \times x^{-1} \right|_m = 1$$

where x and m have no common factor (other than unity). We shall denote this by writing $\left| x^{-1} \right|_m$ for that inverse; that is

$$\left| x \times \left| x^{-1} \right|_m \right|_m$$

In the published literature, the notation $|1/x|_m$ is also frequently used to denote the multiplicative inverse[1]:

$$\left| x \times \left| \frac{1}{x} \right|_m \right|_m = 1$$

So to determine the multiplicative inverse of x, it is sufficient to find a number x^{-1} such that

$$\left| x \times x^{-1} \right|_m = 1 \qquad\qquad 0 \leq x, x^{-1} < m$$

EXAMPLE. Let $x = 7$. To determine the multiplicative inverse of x with respect to 11, we want to find x^{-1} such that

$$\left| 7 \times x^{-1} \right|_{11} = 1$$

That is, we want to find a number whose modular product with 7 is 1. We can readily see that the product of 7 and 8 is 56, and $|56|_{11} = 1$. So the multiplicative inverse of 7 with respect to 11 is 8.
END EXAMPLE

Whereas every residue has an additive inverse, with respect to any modulus, it is not the case that every residue has a multiplicative inverse. Evidently, if the modulus is prime, then every residue, x, with respect to that modulus has a multiplicative inverse. In general, $\left| x^{-1} \right|_m$ exists only if x and m are relatively prime. Table 2.1 shows examples.

---
[1]The reader will also frequently find $x \left| 1/y^{-1} \right|_m$ written as $\left| x/y^{-1} \right|_m$.

Table 2.1: Multiplicative inverses

$m = 7$		$m = 8$	
x	x^{-1}	x	x^{-1}
1	1	1	1
2	4	2	—
3	5	3	3
4	2	4	—
5	3	5	5
6	6	6	—
		7	7

There is no general expression for determining the multiplicative inverse of a number; a brute-force search is about the best one can do. Nevertheless, for prime m, Fermat's Theorem, which is stated next, may sometimes be useful in determining multiplicative inverses.

FERMAT'S THEOREM. For a prime modulus m and a non-negative integer a that is not a multiple of m

$$|a^m|_m = |a|_m$$

PROOF. The proof is by induction on a. The theorem is evidently true for $a = 0$ and for $a = 1$. Consider now the integer $a + 1$. Expanding $(a + 1)^m$, we have

$$(a+1)^m = a^m + C_1^n a^{m-1} + C_1^n a^{m-2} + \cdots + 1 \quad \text{where} \quad C_k^n = \frac{n!}{(n-k)!k!}$$

Except for the first and the last terms, every other term in the expansion has the coefficient m. Therefore, all these terms disappear when a modular reduction is carried out with respect to m. The only terms that then remain are the first and the last ones. So

$$|(a+1)^m|_m = |a^m + 1|_m$$

$$= |\,|a^m|_m + 1|_m \tag{2.3}$$

But, by our hypothesis

$$|a^m|_m = |a|_m \tag{2.4}$$

So substituting from Equation 2.4 into Equation 2.3, we get

$$\big|\,|a^m|_m + 1\big|_m = |a+1|_m$$

whence

$$|(a+1)^m|_m = |a+1|_m$$

Since the theorem is true for $a+1$, by induction it is true for all non-negative a.

END PROOF.

Fermat's theorem is sometimes useful in directly finding the multiplicative inverse. By the theorem

$$|a^m|_m = |a|_m$$

so

$$|a^m a^{-1}| = |aa^{-1}|$$
$$= 1$$

So

$$|a^m a^{-1}|_m = |aa^{m-2}aa^{-1}|_m$$
$$= \big|\,|aa^{-1}|\,|a^{m-2}a|_m\,\big|_m$$
$$= |a^{m-2}a|_m$$

$$= 1$$

from which we may deduce that a and a^{m-2} are multiplicative inverses of each other, with respect to m.

The existence of multiplicative inverses facilitates the solutions, for x, of equations of the form, $|ax|_m = |b|_m$, where a and b are given:

$$|ax|_m = |b|_m$$

$$|axa^{-1}|_m = |bx^{-1}|_m$$

$$|x|_m = |ba^{-1}|_m$$

Division

In conventional arithmetic, division is the most problematic of the basic operations. In residue number systems, there is difficulty in even the seemingly simple matter of defining precisely what division means. It might appear possible to define division in a residue number system by proceeding as we have done above for addition and multiplication—that is, define the operation for residues and then extend that to tuples of residues—but two complications immediately arise if we attempt to do that. The first is that residue division and normal division are in concord only when the quotient resulting from the division is an integer. And the second follows from the fact that zero does not have a multiplicative inverse for zero.

Conventional division may be represented by the equation

$$\frac{x}{y} = q$$

which implies that

$$x = y \times q$$

In residue number systems, however, this last equation does not necessarily hold, since in these systems the fundamental relationship is congruence and not plain equality. Suppose for the RNS equivalent of the preceding equation we take, for residues x and y, the congruence

$$y \times q \equiv x \pmod{m} \tag{2.5}$$

Multiplying both sides by the multiplicative inverse of y, we get

$$q \equiv x \times y^{-1} \pmod{m} \tag{2.6}$$

The proper interpretation of q in these equations is that x is modulo-m sum $\sum_i^q y$. Therefore, q corresponds to the quotient only when q has an integer value. So, unlike the corresponding situation in conventional arithmetic, in residue number systems multiplication by a multiplicative inverse is not always equivalent to division. The following example shows this.

EXAMPLE. Let us assume $m = 5$ and compute the following quotients

$$\frac{4}{2} = q$$

$$2q \equiv 4 \pmod{5}$$
$$q \equiv 4 \times 2^{-1} \pmod{5}$$
$$\equiv 4 \times 3 \pmod{5}$$
$$\equiv 2 \pmod{5}$$

Now consider the cases when integer division is not possible:

$$\frac{4}{3} = q$$

$$3q \equiv 4 \pmod 5$$
$$q \equiv 4 \times 3^{-1} \pmod 5$$
$$\equiv 4 \times 2 \pmod 5$$
$$\equiv 3 \pmod 5$$

and

$$\frac{3}{4} = q$$

$$4q \equiv 3 \pmod 5$$
$$q \equiv 3 \times 4^{-1} \pmod 5$$
$$\equiv 3 \times 4 \pmod 5$$
$$\equiv 2 \pmod 5$$

<u>END EXAMPLE</u>

In all of the three cases in this example, the congruences $q \equiv xy^{-1}$ (mod m) are valid, but q is a correct quotient only in the first case.

The second complication mentioned above arises when we attempt to carry out an extension from residues to tuples of residues, i.e. residue representations of conventional numbers. Suppose that, relative to the moduli $m_1, m_2, \ldots, m_N$, the representations of X and Y are $\langle x_1, x_2, \ldots, x_N \rangle$ and $\langle y_1, y_2, \ldots, y_N \rangle$. Then consider the problem of trying to compute the quotient Q of X and Y. If $Q \cong \langle q_1, q_2, \ldots, q_N \rangle$, then, corresponding to Equations 2.5 and 2.6, we have

$$y_i q_i \equiv x \pmod{m_i}$$
$$q_i \equiv xy_i^{-1} \pmod{m_i} \quad (2.7)$$

For the residue-tuples, we have a modulo-M system, where $M = \prod m_i$; so we seek solutions that satisfy

$$YQ \equiv X \pmod M$$
$$Q \equiv XY^{-1} \pmod M \quad (2.8)$$

Now, even if we accept as solutions for Equation 2.7 only those q_i that correspond to integral solutions (as in the last example), we will not necessarily have solutions to Equation 2.8. The reason for this is that it may be that $y_j = 0$, for some j, and in that case y_j^{-1} does not exist, and neither does Y^{-1}.

From the above, it appears that the sensible way to define division in residue number systems is so that $\langle q_1, q_2, \ldots, q_N \rangle$ is the representation of the result of a conventional division of the conventional numbers X and Y. We may conclude from this that computing the residue-quotient of two residue-numbers (representations of conventional numbers) will require some elements of all three of reverse conversion, conventional division, and forward conversion—rather untidy business.

Scaling

Division in residue number systems is, as we have noted above, a rather difficult operation. Nevertheless, if the divisor is one of the moduli or a product of several moduli, but not a multiple of a power of a modulus, then the division is much easier and is known as *scaling*. Scaling is similar to division by a power of two in a conventional binary system, in which the division is performed by shifting the dividend to the right. Although scaling in residue number systems is not as simple as shifting, it is nevertheless easier to implement than RNS-division by arbitrary numbers. Still, scaling is not without difficulties when it comes to implementation.

Scaling is often useful in preventing overflow, since it reduces the dynamic range of RNS variables and thus keeps them within the permissible bounds. (Recall that detecting overflow is a rather difficult operation in a residue number system.)

Scaling with positive integers may be explained as follows. Consider the division of a number X by a positive integer Y. We may express X as

$$X = \left\lfloor \frac{X}{Y} \right\rfloor \times Y + x$$

The scaling process determines the quotient for certain values of Y. The last equation may be rewritten into

$$\left\lfloor \frac{X}{Y} \right\rfloor = \frac{X - x}{Y} \qquad (2.9)$$

Since we are interested in the obtaining the residue-set of the division operation with respect to $m_1, m_2, m_3 \ldots m_N$, we need to determine the residues of Equation 2.9. For the residue with respect to m_1, we have

$$\left\lfloor \frac{X}{Y} \right\rfloor_{m_1} = \left| \frac{X - x}{Y} \right|_{m_1}$$

By Equation 2.9, these residues will all be integers. Therefore, if Y is a modulus or a product of the first powers of some moduli, then all residues can be determined. With these residues, it is now possible to determine the residues of $\lfloor X/Y \rfloor_{m_i}$:

$$\left\lfloor \frac{X}{Y} \right\rfloor_{m_i} = \left| \frac{X-x}{Y} \right|_{m_i}$$

The remaining residues can be computed through *base extension*, which is briefly discussed in Chapter 1 and in more detail in Chapter 7.

Primitive roots and indices

The main concepts here are those of *indices*, which are similar to logarithms in conventional arithmetic, and *primitive roots*, which are similar to bases of logarithms. Several results are given here without proofs, for which the reader should consult standard texts on elementary number theory (e.g. [8, 9]). We start by recalling of Euler's phi-function:

DEFINITION. For $n \geq 1$, $\phi(n)$ is the number of positive integers less than n and relatively prime to n.

For example, $\phi(3) = 2, \phi(5) = 4, \phi(6) = 2$, and $\phi(30) = 8$. If n is prime, then evidently $\phi(n) = n - 1$. The following theorems can help us compute the values of ϕ without having to resort to brute-force.

THEOREM. If p is a prime integer and $j > 0$, then

$$\phi(p^j) = p^j - p^{j-1}$$
$$= p^j \left(1 - \frac{1}{p}\right)$$

For example, $\phi(16) = \phi(2^4) = 2^4 - 2^3 = 8$. This result can be extended to composite integers as follows.

THEOREM. If m has prime factorization $p_1^{j_1} p_2^{j_2} \cdots p_k^{j_k}$ then

$$\phi(n) = \prod_{i=1}^{k} \left(p^{j_i} - p^{j_i - 1}\right)$$
$$= n \prod_{i=1}^{k} \left(1 - \frac{1}{p_i}\right)$$

For example

$$\phi(360) = \phi(2^3 \times 3^2 \times 5)$$
$$= 360\left(1 - \frac{1}{2}\right)\left(1 - \frac{1}{3}\right)\left(1 - \frac{1}{5}\right)$$
$$= 96.$$

The following are additional useful definitions and results.

DEFINITION. Suppose n and m are relatively prime, and j is the smallest integer such that $n^j \equiv 1 \pmod{m}$. Then j is said to be the *order of* n *modulo* m *(or that* n *is of order j modulo m)*. For example, 3 is the order of 2 modulo 7.

DEFINITION. If n and m are relatively prime and n is order $\phi(n)$ modulo m, then n is a *primitive root* of m. For example, $\phi(7) = 6$, and 3 is a primitive root of 7, because $3^1 \equiv 3 \pmod 7, 3^2 \equiv 2 \pmod 7, 3^3 \equiv 6 \pmod 7, 3^4 \equiv 4 \pmod 7, 3^5 \equiv 5 \pmod 7$.

THEOREM. Suppose n and m are relatively prime and $x_1, x_2, \ldots, x_{\phi(m)}$ are positive integers less than m and relatively prime to m. If n is a primitive root of m, then $n, n^2, \ldots, n^{\phi(m)}$ are congruent (in some order) to $x_1, x_2, \ldots, x_{\phi(m)}$.

COROLLARY. If m has a primitive root, then it has exactly $\phi(\phi(m))$ primitive roots. For example, for $m = 9$, there are $\phi(\phi(9)) = \phi(6) = 2$ primitive roots, which are 2 and 5.

THEOREM. The only integers with primitive roots are 2, 4 and integers of the form m^e or $2m^e$, where m is an odd prime and e is a positive integer.

DEFINITION. If r is a primitive root of m and n and m are relatively prime, then the smallest positive integer j such that $r^j \equiv n \pmod{m}$ is called the *index of n relative to r*. We will use $I(n)$ to denote the index of n.

THEOREM. Suppose m has a primitive root r. Then, for integers x and y

- $I(xy) \equiv I(x) + I(y) \pmod{\phi(m)}$
- $I(x^k) \equiv kI(x) \pmod{\phi(m)}$ for integer $k > 0$
- $I(1) \equiv 0 \pmod{\phi(m)}$
- $I(r) \equiv 1 \pmod{\phi(m)}$

From which we may conclude that if m is chosen appropriately, then in residue number systems operations multiplication and powering can be reduced to log-like addition and multiplication, respectively, followed by appropriate inverse-index (anti-log) operations. This is discussed further in Chapter 5.

2.4 Chinese Remainder Theorem

The Chinese Remainder Theorem (CRT) may rightly be viewed as one of the most important fundamental results in the theory of residue number systems. It is, for example, what assures us that if the moduli of a RNS are chosen appropriately, then each number in the dynamic range will have a unique representation in the RNS and that from such a representation we can determine the number represented. The CRT is also useful in reverse conversion as well as several other operations.

THEOREM. Suppose $m_1, m_2, \ldots, m_N$ are positive pairwise relatively prime integers, that $M = \prod_{i=1}^{N} m_i$, and that $x_1, x_2, \ldots, x_N$ and k are integers. Then there is exactly one integer, X, that satisfies the conditions

$$k \leq X \leq k + M$$
$$x_i = |X|_{m_i} \qquad 1 \leq i \leq N$$

For several constructive and non-constructive proofs, see [5]. We shall give an outline of one of the former in Chapter 7.

An alternative way of stating the same result, in a form we shall later find more useful is as follows. Given a residue representation $\langle x_1, x_2, \ldots, x_N \rangle$, where $x_i = |X|_{m_i}$, the integer represented, X, is given by

$$X = \left| \sum_{i=1}^{N} w_i x_i \right|_M$$

for certain weights, w_i. So, we may expect that constructive proofs will primarily consist of providing information on how to compute the weights.

An important aspect of the second formulation is that it tells us that, if we view the weights as radices, then for a given residue number system, there is a correspondence between RNS representations and representations in some mixed-radix system. We shall return to this point in Chapter 7.

2.5 Complex residue-number systems

We now consider the representation in residue number systems of complex numbers. We cite, without proofs, several results from elementary number theory; for the proofs, the reader should consult appropriate texts, such as [8, 9].

An ordinary residue number system is constructed from a set of distinct pairwise relatively prime integers, $m_1, m_2, \ldots, m_N$, and modular arithmetic operations. A direct approach to the representation of complex RNS-numbers means representing the real and imaginary parts separately. So a starting point in formulating a complex-number RNS (CRNS) is necessary to consider the solutions of the congruence

$$x^2 \equiv -1 \pmod{m_i} \tag{2.10}$$

If this congruence has a solution, j, then -1 is said to be a *quadratic residue*[2] of m_i, if j is an element of the residue-set of m_i. Otherwise, j is a *quadratic non-residue*. Evidently, $j = \sqrt{-1}$ is a solution of Equation 2.10, but it is nnot necessarily congruent to an element in the residue-set of m_i and so is a quadratic non-residue. The basic properties of congruences show that if j corresponds to a quadratic residue, then so does $m_i - j$.

EXAMPLE. Take $m_i = 5$, then -1 is a quadratic residue of 5 because

$$2^2 = 4$$
$$\equiv -1 \pmod{5}$$

On the other hand, if $m_i = 7$, then Equation 2.10 has no solution, as can be ascertained by examining $x = 1, 2, 3, 4, 5, 6$. Therefore -1 is a non-quadratic residue of 7. END EXAMPLE

To determine when Equation 2.10 has solutions and what type of solutions they are, the following results from elementary number theory are useful.

THEOREM. -1 is a quadtratic residue if p is a prime of the form $4k+1$ and a quadratic non-residue if p is a prime of the form $4k+3$.

If m_i is not prime, then we first decompose it into its prime factors. A quadratic residue then exists if each prime factor is of the form $4k+1$.

---

[2] In general, we are looking for residues that are solutions to the quadratic congruence $ax^x + bx + c \equiv 0 \pmod{m}$.

Regardless of whether a solution, j, is a quadratic residue or a quadratic non-residue, we can form a CRNS, by taking ordered pairs of residues, (x_R, x_I), where x_R and x_I are in the residue-set of the given modulus:
$$(x_R, x_I) \Leftrightarrow x_R + jx_I$$
The primary difference between the two types of solution to Equation 2.10 is that for a quadratic residue, $x_R + jx_I$ will be real, whereas for a non-quadratic residue, it will have an imaginary component.

Addition/subtraction and multiplication in a CNRS are defined in a manner similar to ordinary complex-number arithmetic. Thus, for addition and subtraction we have
$$(x_R, x_I) \pm (y_R, y_I) \stackrel{\triangle}{=} (x_R + jx_I) \pm (y_R + jy_I)$$

$$= |x_R + y_R|_{m_i} \pm j\,|x_I + y_I|_{m_i})$$

$$\stackrel{\triangle}{=} \left(|x_R + y_R|_{m_i},\, |x_I + y_I|_{m_i}\right)$$
provided, $(x_R, x_I, y_R,$ and y_I are in the residue-set of m_i. Similarly, multiplication, which involves involves four cross-products, as in conventional complex arithmetic, is defined as
$$(x_R, x_I) \times (y_R, y_I) \stackrel{\triangle}{=} (x_R + jx_I) \times (y_R + jy_I)$$

$$= |x_R y_R - x_I y_I|_{m_i} + j\,|x_I y_I + x_R y_I|_{m_i}$$

$$\stackrel{\triangle}{=} \left(|x_R y_R - x_I y_I|_{m_i},\, |x_I y_I + x_R y_I|_{m_i}\right)$$
We have thus far assumed that m_i is prime. If m_i is not prime, the above formulation of a CRNS still goes through, regardless of the solutions of Equation 2.10.

It is possible to formulate a complex-number system — a quadratic RNS (QRNS) — that is, in some ways, better than the one outlined above, if Equation 2.10 has a solution that is a quadratic residue. To do so, map between the tuples (x_R, x_I) and the tuples (X, X^*):

$$(x_R, x_I) \Leftrightarrow (X, X^*)$$
where
$$X = |x_R + jx_I|_{m_i}$$

$$X^* = |x_R - jx_I|_{m_i}$$

and, therefore,
$$x_R = \left| \frac{|X + X^*|_{m_i}}{2} \right|_{m_i}$$

$$x_I = \left| \frac{|X - X^*|_{m_i}}{2j} \right|_{m_i}$$

These last two equations can be simplified by using multiplicative inverses:

$$x_R = \left| |2^{-1}|_{m_i} |X + X^*|_{m_i} \right|_{m_i}$$

$$x_I = \left| |2^{-1}|_{m_i} |j^{-1}|_{m_i} |X + X^*|_{m_i} \right|_{m_i}$$

where X and X^* are real numbers and $|j^{-1}|_{m_i}$ are the multiplicative inverses of 2 and j with respect to m_i.

The arithmetic operations in this system are similar to those in CRNS. Addition and multiplication in the QRNS are defined as

$$|(X, X^*) \pm (Y, Y^*)|_{m_i} = \left(|X \pm Y|_{m_i}, |X^* \pm Y^*|_{m_i} \right)$$

$$|(X, X^*) \times (Y, Y^*)|_{m_i} = \left(|X \times Y|_{m_i}, |X^* \times Y^*|_{m_i} \right)$$

From which we observe that in QRNS multiplication does not involve cross-product terms as in ordinary complex multiplication or in CRNS. This yields several advantages, relative to CRNS: hardware will be simpler and more regular, higher hardware performance is possible, an error in some digit has no effect on other digits or other results, and so forth. There is, however, some overhead in the conversion between RNS and QRNS.

2.6 Redundant residue number systems

As we remarked in Chapter 1, one of the main advantages of residue number systems is that they facilitate the detection and correction of errors. This arises from the fact that in the residue representation of a number, all the digits are independent; therefore, an error in one digit-position does not corrupt any other digit-position. So if an error occurs in some digit-position, computations may still proceed, through the exclusion of the faulty digit-position (and corresponding modulus), provided that either the resulting smaller dynamic range is acceptable, or that the original system had some extra moduli that provided a larger range than that nominally required for the computations to be carried out. Note, though, that while fail-soft capability exists for all operations, error-isolation is not possible with operations that require interaction between all digits of an RNS representation; that is, operations such as division, magnitude-comparison, and reverse conversion.

Suppose we have determined that the N moduli $m_1, m_2, \ldots, m_N$ are sufficient to provide the dynamic range that we require. If we then add R extra moduli, $m_{N+1}, m_{N+2}, \ldots, m_{N+R}$, then we have a *redundant residue number system*, and these extra moduli are the *redundant moduli*. We thus have two dynamic ranges: $M \triangleq \prod_{i=1}^{N}$ and $M_R \triangleq \prod_{i=1}^{N+R}$. M defines the *legitimate range* of the computations, and the extra $M_R - M$ states constitute the *illegitimate range*. An error is known to have occurred if a result is within the latter range. If $r \geq 2$, then an error in a single digit can be corrected [6, 7]. In general, R redundant moduli pert the detection of up to R errors and the correction of up to $R/2$ errors. Overflow can also be detected by determining the range in which the result lies, although this is by no means an easy operation.

If negative numbers are permitted, then the mapping given in Chapter 1 shows that positive numbers will be mapped onto $[0, (M-1)/2]$ if M is odd, or to $[0, M/2 - 1]$ if M is even, and negative numbers will be mapped onto $[M_R - (M-2)/2, M_R - 1]$ if M is odd, or to $[M_R - M/2, M_R - 1]$ if M is even. The negative numbers therefore fall into the illegitimate range and need to be brought back into the proper dynamic range. This can be achieved by applying a circular shift (also known here as a *polarity shift*) that effectively adds, to every number, $(M+1)/2$ if M is odd, or $M/2$ if M is even.

Note that in a redundant RNS, numbers are still nominally represented only within the underlying non-redundant system; however, in the computations the redundant system must be used. An obvious question then arises of how to compute the redundant residues given the non-redundant ones. This can be done through *base extension*, which is discussed in detail in Chapter 7.

A redundant RNS may be used as follows for error correction. The redundant digits of a representation are used to access a look-up table (LUT) that contains the corresponding correction; the contents of this LUT can be constructed by using base-extension. Alternatively, the use of a LUT may be replaced with extra computations.

2.7 The Core Function

We noted in Chapter 1 that certain operations are fundamentally difficult in RNS; these include magnitude-comparison, sign-determination, overflow-determination, and division. The essence of the difficulties is that RNS digits carry no weight information, and therefore it is not easy to determine the magnitude of a given RNS number. The Core Function [2] provides some means that can help deal with the difficult operations; briefly, one may view it as a method for approximating the magnitude of an RNS number. The underlying idea is to map an RNS number $n \in [0, M-1]$ to a number $C(n) \in [0, C(M)]$, where $C(M) << M$ and M is the product of the moduli $m_1, m_2, \ldots, m_N$ in the RNS. Applications of the Core Function are discussed in Chapters 6 and 7.

Let $\langle x_1, x_2, \cdots, x_N \rangle$ be the residue representation of x, with respect to the moduli $m_1, m_2, \cdots, m_N$. Then the *core*, $C(x)$, of x is defined by

$$C(n) = \sum_{i=1}^{N} \left\lfloor \frac{x}{m_i} \right\rfloor$$

$$= \sum_{i=1}^{N} \frac{w_i}{m_i} n - \sum_{i=1}^{N} \frac{w_i}{m_i} |x|_{m_i} \qquad (2.11)$$

where w_i are weights that are determined in a manner to be described. So

$$C(M) = \sum_{i=1}^{N} w_i \frac{M}{m_i} - \sum_{w_i} \frac{|M|_{m_i}}{m_i}$$

The second term is evidently zero, since M (by definition) is exactly divisible by m_i, so

$$C(M) = \sum_{i=1}^{N} w_i \frac{M}{m_i} \qquad (2.12)$$

and

$$\frac{C(M)}{M} = \sum_{i=1}^{N} \frac{w_i}{m_i}$$

Substituting for $\sum w_i / m_i$ in Equation 2.7, we get

$$C(x) = \frac{C(M)}{M} x \sum_{i=1}^{N} \frac{w_i}{m_i} x_i \qquad (2.13)$$

Equation 2.13 shows that we may view the Core Function as consisting of a linear function (the first term) to which some noise (the second term) has been added.[3] Ideally, we should like the noisy part to be (relative to the first part) as small as possible—the smaller the noise, the better the function is in giving us an idea of the relative magnitude of given RNS numbers. Thus, $C(M)$ should be as large as possible, or the weights should be as small as possible. There are, however, some constraints: To keep hardware simple and balanced, $C(M)$ should be of roughly the same magnitude as the moduli, and these are likely to be rather small. Choices for the weights are also restricted by the solutions to the equations above. One way in which the noisy part can be readily reduced is by replacing the definition above of the Core Function (Equation 2.11) with [3]

$$C(x) = \frac{C(M)}{M}x - \sum_{i=1}^{N} \frac{|w_i x_i|_{m_i}}{m_i}$$

but this version is not conducive to certain algebraic manipulations.

The core of a number can be obtained by using a Chinese Remainder Theorem for Core Functions, but this sometimes produces values that have some ambiguity. This ambiguity can be eliminated by the use of an extra modulus, m_E, that is greater than the difference between the smallest and largest values assumed by the given core function. Reduction, with respect to m_E, of both sides of Equation 2.13 then yields

$$|C(x)|_{m_E} = \left| \frac{C(M)}{M}x - \sum_{i=1}^{N} \frac{w_i}{m_i}x_i \right|_{m_E}$$

$$= \left| \frac{C(M)}{M}|x|_{m_E} - \sum_{i=1}^{N} \frac{w_i}{m_i}x_i \right|_{m_E}$$

Ambiguities are then eliminated, because the core is evaluated relative a modulus larger than its range. To avoid complex computations for $|x|_{m_E}$, [4] has instead proposed the use of a "parity" bit; this requires that M (and therefore each moduli) be odd. Essentially, the use of a parity bit amounts to taking m_E to be as small as possible and without any constraints: $m_E =$

---

[3] Another way to view the "noise" is that it captures the "fuzziness" in how finely we can differentiate between two residue numbers.

2. Thus the parity bit, $p(x)$, of x is defined to be $|x|_2$. So we now have

$$|C(x)|_{m_E} = \left| \frac{C(M)}{M} x - \sum_{i=1}^{N} \frac{w_i}{m_i} x_i \right|_2$$

$$= \left| \frac{C(M)}{M} |x|_2 - \sum_{i=1}^{N} \frac{w_i}{m_i} x_i \right|_2$$

$$= \left| p + \sum_{i=1}^{N} w_i x_i \right|_2 \qquad \text{since } C(M) \text{ and } m_i \text{ are odd}$$

The weights can be found as follows. Let $M_i \triangleq M/m_i$. Then from Equation 2.12, we have

$$|C(M)|_{m_i} = \left| \sum_{i=1}^{N} w_i M_i \right|_{m_i}$$

$$= |w_i|_{m_i} \qquad (2.14)$$

since all but one of the terms in the sum is a multiple of m_i. Therefore

$$|w_i|_{m_i} = \left| C(M) \left| M_i^{-1} \right|_{m_i} \right|_{m_i}$$

where $\left| M_i^{-1} \right|_{m_i}$ is the multiplicative inverse of M_i with respect to m_i. Thus if $C(M)$ is given, then the w_i are obtainable.

EXAMPLE. Take the moduli-set $\{3, 5, 7\}$, and let $C(M)$ be 31. Then

$$M_1 = 35 \qquad \left| M_1^{-1} \right|_3 = 2$$

$$M_2 = 21 \qquad \left| M_2^{-1} \right|_5 = 1$$

$$M_3 = 15 \qquad \left| M_3^{-1} \right|_7 = 1$$

from which we get, by Equation 2.14

$$w_1 = |31 \times 2|_3$$
$$= 2 \text{ or } -1$$
$$w_2 = |31 \times 1|_5$$
$$= 1 \text{ or } -4$$
$$w_3 = |31 \times 1|_7$$
$$= 3 \text{ or } -4$$

We want the "noise" in the core function to be as small as possible, so we pick those values of weights that have the smallest magnitudes: $w_1 = -1$, $w_2 = 1$, $w_3 = 3$.

END EXAMPLE

2.8 Summary

This chapter has introduced the mathematical foundations of residue number systems. Most of the results introduced, even the rather elementary ones, will be very useful for what is in subsequent chapters. The more advanced results, such as the Chinese Remainder Theorem and the Core Function, have very significant uses, especially with respect to difficult RNS operations, e.g. division and conversion from residue to conventional notations. Redundant residue number systems have also been introduced; these, and their applications, are are discussed in more detail in Chapter 8.

References

(1) N.S. Szabo and R. I. Tanaka. 1967. *Residue Arithmetic and Its Applications to Computer Technology*. McGraw-Hill, New York.
(2) D.D. Miller. 1986. Analysis of the residue class core function of Akushskii, Burcev, and Park. In: G. Jullien, Ed., *RNS Arithmetic: Modern Applications in Digital Signal Processing*. IEEE Press.
(3) J. Gonnella. 1991. The application of the core function to residue number systems. *IEEE Transactions on Signal Processing*, SP-39:69–75.
(4) N. Burgess. 1997. "Scaled and unscaled residue to binary number conversion techniques using the core function". In: *Proceedings, 13th International Symposium on Computer Arithmetic*, pp 250–257.

(5) D. E. Knuth. 1969. *The Art of Computer Programming, Vol. 2*. Addison-Wesley, Reading, MA.
(6) H. Krishna and J.-D. Sun. 1993. On theory and fast algorithms for error correction in residue number system product codes. *IEEE Transactions on Computers*, 42(7):840–852.
(7) C.-G. Sun and H.-Y. Lo. 1990. An algorithm for scaling and single error correction in residue number systems. *IEEE Transactions on Computers*, 39(5):1053–1064.
(8) G. H. Hardy and E. M. Wright. 1979. *An Introduction to the Theory of Numbers*. Oxford University Press, UK.
(9) D. M. Burton. 1980. *Elementary Number Theory*. Allyn and Bacon, Boston, USA.

Chapter 3

Forward conversion

Numbers that are initial inputs to, or final outputs from, residue computations will usually be in some conventional notation. *Forward conversion* is the process of translating from conventional notation, here binary or decimal, into residue notation. This chapter covers algorithms and architectures for that process. We shall divide our discussion according to two classes of moduli-sets: arbitrary sets and the special moduli-sets (i.e. those of the form $\{2^n - 1, 2^n, 2^n + 1\}$ and extensions thereof). The basic principle in the computation of residues is division, with the moduli as the divisors. But division is an expensive operation in hardware and so is rarely used in the computation of residues, whence the significance of the special moduli. Division is avoidable in the case of other moduli as well, but the hardware required will not be as simple as in the case of the special moduli.

Hardware implementations for forward conversion may be based on look-up tables (the use of which is facilitated by the small sizes of typical moduli), combinational-logic circuits, or a mixture of both. Converters for the special moduli-sets are almost always implemented in combinational logic, whereas those for arbitrary sets will be of any of the three types. The complexity of the conversion depends on the moduli-set chosen for a specific application. Signal processing applications tend to require a large dynamic range, and the moduli-sets used in these cases will consist of either a large number of small relatively-prime numbers or of a small number of large relatively-prime numbers; more often, it will be the latter. The amount of memory used in these converters is generally proportional to both the magnitude of the numbers involved as well as the number of moduli in the set used. Although forward converters for the special moduli-sets are easily implemented, for large dynamic ranges, large moduli may be necessary, which can necessitate the use of complex processing units and thus offset the advantages of such moduli-sets in particular and of residue number systems in general.

The chapter consists of three main sections. The first section deals with converters for the special moduli-sets. The second deals with memory-based converters for arbitrary moduli-sets. And the third covers combinational-logic converters for arbitrary moduli-sets. In reading what follows, the reader may find it useful to periodically refer back to Chapter 2.

3.1 Special moduli-sets

In this section we shall look at forward conversion for the special moduli, $2^n - 1, 2^n$, and $2^n + 1$. Given that arithmetic modulo-2^n is just conventional arithmetic, what follows covers only the moduli $2^n - 1$ and $2^n + 1$. The special moduli are usually referred to as *low-cost moduli*, since conversion to and from their residues can be realized relatively easily and does not require complex operations, such as evaluation of multiplicative inverses, multiplication, and so forth [7].

Consider the computation of the residue of an arbitrary integer X with respect to a modulus m. Since X may be represented as an n-bit binary number, $x_{n-1}x_{n-2}\cdots x_0$, and its residue with respect to m may be expressed as

$$|X|_m = |x_{n-1}x_{n-2}x_{n-3}\cdots x_0|_m$$

of which an equivalent expression is

$$|X|_m = \left|2^{n-1}x_{n-1} + 2^{n-1}x_{n-2} + 2^{n-1}x_{n-3}\ldots\ldots + 2^{n-1}x_0\right|_m$$

From the properties of residues given in Chapter 2, we have

$$|X|_m = \left|\,\left|2^{n-1}x_{n-1}\right|_m + \left|2^{n-2}x_{n-2}\right|_m + \left|2^{n-3}x_{n-3}\right|_m + \ldots + \left|2^0 x_0\right|_m\right|_m$$

Since x_i is either 0 or 1, in computing the residue of X, all that is required is the evaluation of the values $|2^i|_m$, which are then added up, with a reduction relative to the modulus.

Modulus $2^n - 1$

Conversion and arithmetic modulo $2^n - 1$ are quite easy to implement. For example, relative to conventional (i.e. two's complement) binary addition, the only "complication" in modulo-$(2^n - 1)$ addition is the necessity to sometimes also add an end-around-carry.[1] So simple hardware circuits can be realized with this category of modulus [2].

The residues with respect to $2^n - 1$ are determined as follows. Observe that

$$|2^n|_{2^n-1} = |2^n - 1 + 1|_{2^n-1} \qquad (3.1)$$
$$= 1$$

and that this equation can be extended easily to a product in the exponent, 2^{nq}, and, in general, to an arbitrary power of 2:

$$|2^{nq}|_{2^n-1} = \left| \prod_{i=1}^{q} |2^n|_{2^n-1} \right|_{2^n-1} \qquad (3.2)$$
$$= 1 \qquad \text{(by Equation 3.1)}$$

So the residue of any number 2^m, where $m \neq n$, can be determined by using Equations 3.1 and 3.2:

$$|2^m|_{2^n-1} = |2^{nq+r}|_{2^n-1}$$

$$= \left| \, |2^{nq}|_{2^n-1} \times |2^r|_{2^n-1} \right|_{2^n-1}$$

$$= 1 \times |2^r|_{2^n-1} \qquad (3.3)$$

where $q = \lfloor m/n \rfloor$, and r is the remainder from the division. The following example shows such an evaluation.

EXAMPLE. Take $X = 2^7$ and $m = 2^3 - 1$ (i.e. $n = 3$ and $m = 7$). Then

$$|X|_7 = |2^7|_7$$

$$= \left| \, |2^{2 \times 3}|_7 \times |2|_7 \right|_7$$

In this example, $q = 2$ and $r = 1$. Therefore, by Equation 3.2,

$$|X|_7 = |1 \times 2|_7$$
$$= 2$$

END EXAMPLE

We next show that a similar approach can be used in the computation of residues with respect to the other low-cost modulus, $2^n + 1$.

---

[1] Note that this is just one's complement addition.

Modulus $2^n + 1$

As in the preceding case, we begin by considering the residue of 2^n with respect to the modulus $2^n + 1$:

$$|2^n|_{2^n+1} = |2^n + 1 - 1|_{2^n+1}$$
$$= -1$$

We then extend this to an arbitrary power of two, 2^m, where $m \neq n$ and $m = nq + r$:

$$|2^m|_{2^n+1} = |2^{nq}|_{2^n+1} \times |2^r|_{2^n+1}$$

$$= \begin{cases} 2^r & \text{if } q \text{ is even} \\ 2^n + 1 - 2^r & \text{otherwise} \end{cases}$$

where $q = \lceil m/n \rceil$. When q is odd, $|2^{nq}|_{2^n+1}$ is -1, and so $2^n + 1$ must be added back to make the residue positive.

EXAMPLE. Take $X = 2^7$ and $m = 2^3 + 1$; i.e. $n = 3$ and $m = 9$. Then q is even, and so

$$|2^7|_9 = |2^{3 \times 2}|_9 \times |2|_9$$
$$= 2$$

Now, take $X = 2^7$ and $m = 2^2 + 1$; i.e. $n = 2$ and $m = 5$. Then q is odd, and

$$|2^7|_5 = |2^{2 \times 3}|_5 \times |2|_5$$
$$= 2^2 + 1 - 2$$
$$= 3$$

END EXAMPLE

3.1.1 $\{2^{n-1}, 2^n, 2^{n+1}\}$ *moduli-sets*

We have seen above how to obtain the residues relative to each of the moduli $2^n + 1$ and $2^n - 1$. The only other modulus in the basic special set is 2^n. Residues with respect to this modulus are obtained easily by dividing the given binary number by 2^n, which "division" is just an n-bit right-shift of the given binary number, X. So forward conversion in the $\{2^n - 1, 2^n, 2^n + 1\}$ moduli-set is straightforward and simple logic circuits, involving modular adders, will suffice for the implementation. If we define,

$m_1 \triangleq 2^n + 1, m_2 \triangleq 2^n$ and $m_3 \triangleq 2^n - 1$, then any integer X within the dynamic range, $M \triangleq [0, 2^{3n} - 2^n - 1]$ (where the upper end of the range is $m_3 m_2 m_1$), is uniquely defined by a residue-set $\{r_1, r_2, r_3\}$, where $r_i = |X|_{m_i}$ and X is a $3n$-bit:

$$X = x_{3n-1} x_{3n-2} \ldots x_{2n} x_{2n-1} \ldots \ldots x_n x_{n-1} \ldots \ldots x_0$$

Residues are obtained by nominally dividing X by m_i. The residue r_2 is the easiest to compute: The n least significant bits constitute the remainder when X is divided by 2^n. Hence r_2 is the number represented by the least significant n bits of X. These bits are obtained by nominally shifting to the right by n bits; "nominally" because the shift may be hardwired.

In order to determine the residues, r_1 and r_3, we first partition X into three n-bit blocks, $\mathbf{B}_1, \mathbf{B}_2, \mathbf{B}_3$ [12,13]:

$$\mathbf{B}_1 \triangleq \sum_{j=2n}^{3n-1} x_j 2^{j-2n}$$

$$\mathbf{B}_2 \triangleq \sum_{j=n}^{2n-1} x_j 2^{j-n}$$

$$\mathbf{B}_3 \triangleq \sum_{j=0}^{n-1} x_j 2^j$$

Then

$$X = \mathbf{B}_1 2^{2n} + \mathbf{B}_2 2^n + \mathbf{B}_3$$

The residue r_1 is then obtained as

$$r_1 = |X|_{2^n+1}$$

$$= \left| \mathbf{B}_1 2^{2n} + \mathbf{B}_2 2^n + \mathbf{B}_3 \right|_{2^n+1}$$

$$= \left| |\mathbf{B}_1 2^{2n}|_{2^n+1} + |\mathbf{B}_2 2^n|_{2^n+1} |\mathbf{B}_3|_{2^n+1} \right|_{2^n+1}$$

$\mathbf{B}_3$ is an n-bit number and therefore is always less than $2^n + 1$; so its residue is simply the binary equivalent of this term. The residues of the other two sums are computed as

$$\left| \mathbf{B}_1 2^{2n} \right|_{2^n+1} = \left| |\mathbf{B}_1|_{2^n+1} |2^{2n}|_{2^n+1} \right|_{2^n+1}$$

and
$$|\mathbf{B}_2 2^n|_{2^n+1} = ||\mathbf{B}_2|_{2^n+1} |2^n|_{2^n+1}|_{2^n+1}$$

Each of $\mathbf{B}_1$ and $\mathbf{B}_2$ is represented in n bits and there must be less than $2^n + 1$. And the residue of 2^{2n} with respect to $2^n + 1$ is

$$\begin{aligned}|2^{2n}|_{2^n-1} &= |2^n 2^n|_{2^n+1} \\ &= |2^n + 1 - 1|_{2^n+1}|2^n + 1 - 1|_{2^n+1} \\ &= -1 \times -1 \\ &= 1\end{aligned}$$

It follows from this that the residue of 2^n with respect to $2^n + 1$ is -1. Therefore,

$$r_1 = |\mathbf{B}_1 - \mathbf{B}_2 + \mathbf{B}_3|_{2^n+1} \qquad (3.4)$$

Similarly, to compute r_3, we first observe that

$$\begin{aligned}|2^{2n}|_{2^n-1} &= |2^n - 1 + 1|_{2^n-1} \times |2^n - 1 + 1|_{2^n-1} \\ &= 1 \times 1 \\ &= 1\end{aligned}$$

Also, $|2^n|_{2^n-1}$ is 1. So

$$r_3 = |\mathbf{B}_1 + \mathbf{B}_2 + \mathbf{B}_3|_{2^n-1} \qquad (3.5)$$

From the above, we may surmise three modular adders will suffice for the computation of the residues. If the magnitudes of the numbers involved are small, as will be the case for small moduli, the complexity of the overall conversion will not be high. We illustrate the conversion with the following example.

EXAMPLE. Consider the moduli-set $\{7, 8, 9\}$, and let $X = 167$. The binary representation of X is 10100111. Since $n = 3$ in the given moduli-set, we partition X into 3-bit blocks, starting from the right: $\mathbf{B}_1 \triangleq 010$ $\mathbf{B}_2 \triangleq 100$ $\mathbf{B}_3 \triangleq 111$. Applying Equation 3.4, we get (in decimal)

$$\begin{aligned}|167|_{2^3+1} &= |167|_9 \\ &= |2 - 4 + 7|_9 \\ &= 5\end{aligned}$$

The residue with respect to 8, which 2^3, is obtained by shifting the binary equivalent of 167 three bits to the right and taking the three bits shifted out: 7 in decimal. Finally, for the residue with respect to 7, which is $2^3 - 1$, we have, by Equation 3.5

$$|167|_{2^3-1} = |167|_7$$
$$= |2 + 4 + 7|_7$$
$$= 6$$

Hence the residue representation of X is $\langle 5, 7, 6 \rangle$.
END EXAMPLE

Implementation

Figure 3.1 shows the organization of a basic unit for forward conversion as described above. The modulo adders may be realized as all-ROM, all-combinational-logic, or a combination of both; we shall here assume pure combinational logic. The implementation of modular adders is discussed in detail in Chapter 4.

Although the design of Figure 3.1 is straightforward, the modular adders must be full carry-propagate adders[2], and this may result in low performance. There are several ways in which performance can be improved, but all will require some extra logic.

There are two fundamental techniques that may be used here to attain better performance: the first is the use of carry-save adders[3]; and the other is the exploitation of more parallelism. Figure 3.2 shows a design based on these techniques. A carry-save adder (CSA) takes three operands and produces two outputs, a partial-sum (PS) and a partial-carry (PC), that finally must be fed into a carry-propagate adder (CPA) so that the carries are propagated to produce a result in conventional form. The computation of r_1 may require a corrective subtraction of the modulus m_3, and that of r_3 may require the subtraction m_1 or $2m_1$. All the different possible results are computed in parallel and the correct one then selected through a multiplexer. It should be noted that although separate adders are shown in Figure 3.2, in practice the adder-logic can be shared; so the replication

---

[2] In a carry-propagate adder, carries are propagated between digits. The addition time is therefore a function of the number of digits—for n digits, the operational time will be between $O(n)$ and $O(\log n)$.

[3] A carry-save adder consists of just a sequence of full adders. The addition time is therefore constant and independent of the number of digits, but at some point carries must still be propagated. The advantage comes from the fact that in a series of additions, the carries may be saved for propagation at the last addition.

will be less than appears at first glance. Also, for certain high-speed CPA-designs, and depending on the precision of the moduli, the differences in performance between Figure 3.1 and Figure 3.2 may not be substantial.

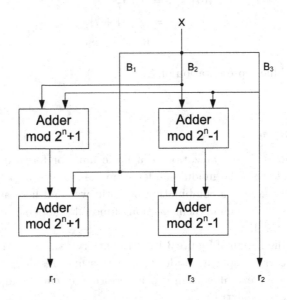

Figure 3.1: $\{2^n - 1, 2^n, 2^n + 1\}$ forward converter

3.1.2 Extended special moduli-sets

The method described above can be applied equally well to derivatives of the 2^n-moduli-sets. We shall here consider just two example moduli-sets in this category: those of the form $\{2^n - 1, 2^n, 2^n + 1, 2^{n+1} - 1\}$, for odd n, and those of the form $\{2^n - 1, 2^n, 2^n + 1, 2^{n+1} + 1\}$ for even n. These are frequently encountered in applications where a large dynamic range is required. We shall adopt essentially the same procedure described above; but in the computation of the residue corresponding to 2^{n+1}, the binary representation of X will now be partitioned into blocks of $n + 1$ bits each. The following example illustrates the procedures used to compute the residues for each of the two sets [5].

EXAMPLE. Take the moduli-set $\{3, 4, 5, 7\}$. The moduli in this set are relatively prime, and the fourth modulus is of the form $2^{n+1} - 1$, with $n = 2$. Let X be 319. In order to obtain the residues with respect to 3,4, and 5, we proceed exactly as in the last example above. We partition the binary representation of 319, which is 0100111111, into 2-bit blocks, since $n = 2$:

$$|319|_3 = |01 + 00 + 11 + 11 + 11|_{2^2-1}$$
$$= |1 + 0 + 3 + 3 + 3|_3$$
$$= 1$$

Similarly, the residue with respect to 5 is obtained as

$$|319|_5 = |01 + 00 + 11 + 11 + 11|_{2^2+1}$$
$$= |1 - 0 + 3 - 3 + 3|_5$$
$$= 4$$

END EXAMPLE

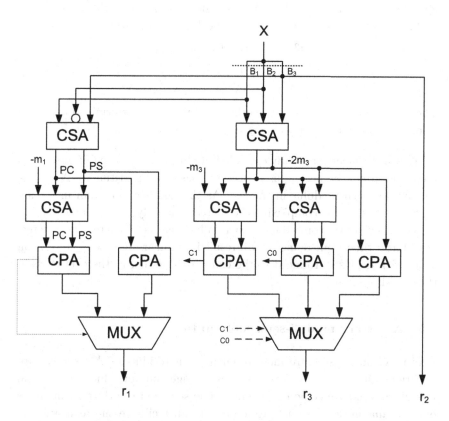

Figure 3.2: Enhanced-performance $\{2^n - 1, 2^n, 2^n + 1\}$ forward converter

The residue with respect to 4 is obtained by simply right-shifting the binary equivalent by two bit-positions, with 3 as the value of the bits shifted out. In order to determine the residue with respect to 7, we partition the binary representation of 319 into 3-bit blocks because the binary representation of 7 is in three bits:

$$|319|_7 = |100\ 111\ 111|_{2^3-1}$$
$$= |4 + 7 + 7|_7$$
$$= 4$$

END EXAMPLE

Let us now consider a similar case but with n odd.

EXAMPLE. Let $n = 3$. In this case the moduli-set is the form $\{2^n - 1, 2^n, 2^n + 1, 2^{n+1} + 1\}$; that is, it will be $\{7, 8, 9, 17\}$. And

$$|319|_{17} = |1\ 0011\ 1111|_{17}$$
$$= |1 - 3 + 15|_{17}$$
$$= 13$$

The residues with respect to the other moduli can be determined as shown in preceding examples. END EXAMPLE

Forward conversion in any extended moduli-set consisting of more than four moduli can be accomplished equally easily by partitioning X appropriately for each modulus in the set. As the number of moduli in the set increases, the circuit-complexity will increase linearly. The complexity can be reduced if conversion is done sequentially—that is, if one residue is determined at a time—but doing so will result in a linear increase in conversion time, with delays introduced in the converters adding to the overheads in the overall system.

3.2 Arbitrary moduli-sets: look-up tables

Arbitrary moduli-sets are most often used in residue-number-system applications that require a large dynamic range but for which the special moduli-sets impose some constraints. The selection of arbitrary numbers of small magnitude for such a moduli-set can facilitate the realization of simple processing elements in subsequent processing, but the forward conversion will still not be as simple as in the special cases, since it cannot be accomplished by a straightforward partitioning of the operand [9,10]. So forward conversions with arbitrary moduli-sets tend to be complex and require elaborate hardware. The complexity of the conversion here will be a function of both the number of moduli employed and the magnitude of each modulus.

In principle, look-up tables (typically implemented as ROM) can be employed directly by having tables that store all possible residues and which are addressed by the numbers whose residues are required [6]. Evidently, the amount of memory necessary can be quite large. Nevertheless, the basic idea is of practical value when combined with techniques that reduce the basic problem to a smaller one or which exploit certain properties of residues.

The basic idea in the use of look-up tables is that finding the residue of a number with respect to a given modulus essentially boils down to computing the modular sum of certain powers of two. Suppose we have an n-bit number, X, and that we wish to compute its residue with respect to a modulus m:

$$X \triangleq x_{n-1}x_{n-2}x_{n-3}.........x_0$$
$$= \sum_{j=0}^{n-1} x_j 2^j$$

and

$$|X|_m = \left| \sum_{j=0}^{n-1} x_j 2^j \right|_m$$
$$= \left| \sum_{j=0}^{n-1} |x_j 2^j|_m \right|_m \quad (3.6)$$

There are many ways in which the partial sums and the total sum may be computed—serially, sequentially, in parallel, or some other combination of these—and these give rise to a variety of architectures.

3.2.1 Serial/sequential conversion

A simple and direct way to implement Equation 3.6 is to have a sequential structure that consists of a look-up table that stores all the values $|2^j|_m$ (note that x_j is either 0 or 1), a modular adder that computes partial sums, a counter, and an accumulator register. The basic structure is shown in Figure 3.3 and operates as follows. The accumulator is initialized to zero. Thereafter, on each cycle, a value is read from the look-up table and added to the value in the accumulator. The process will take n cycles, but the hardware requirements are low: a shift-register that in each cycle makes available one bit of X, a counter whose contents (from 0 to $n-1$) are used in each cycle to address the look-up table, a memory of size about $n \log_2 m$ bits, and an m-bit modular adder. This, though, is not as low as one can get down the cost: for a minimal-cost design, all the basic units in Figure 3.3 may be implemented to work bit-serially.

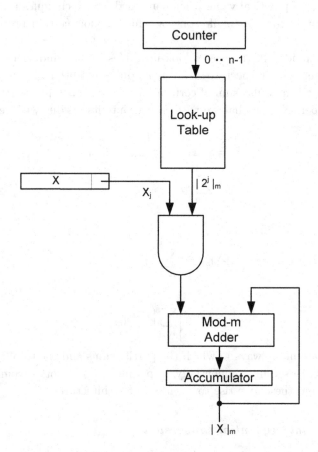

Figure 3.3: Sequential table-lookup converter

The primary drawback of the design in Figure 3.3 is the obviously low performance. Depending on the relative costs and speed of memory and adders, one could modify this slightly to the structure shown in Figure 3.4. Here, the operand, X, is processed two bits at a time. That is, in each cycle, two values, $\left|x_j 2^j\right|_m$ and $\left|x_{j+1} 2^{j+1}\right|_m$ are added and the result then added to that in the accumulator. Some performance-enhancing pipelining is possible with this design.

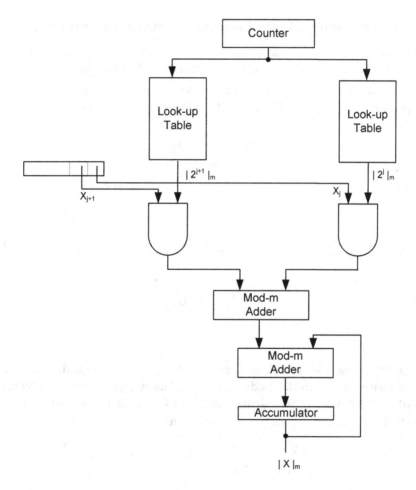

Figure 3.4: Modified sequential table-lookup converter

The move from Figure 3.3 to Figure 3.4 involves a fundamental technique often employed in the quest for high-speed arithmetic units—work with more bits at a time. Thus instead of one or two bits of X, we may take any number of bits and have a completely parallel structure or a sequential-parallel one. Essentially, if we view the basic approach outlined above as being a radix-2 computation, then we now seek to compute in a higher radix—2^R for R bits at a time. We next consider how this can be done.

3.2.2 Sequential/parallel conversion: arbitrary partitioning

Suppose that instead of taking the bits of X serially (i.e. one bit at a time) we instead take p bits at a time [10]. In this case, X is now partitioned into $k \triangleq n/p$ blocks of p bits each. (Without loss of generality, we assume that n/p is integral.) Let the blocks be $\mathbf{B}_{k-1}, \mathbf{B}_{k-2}, \ldots, \mathbf{B}_0$. Then

$$X = \sum_{j=0}^{k-1} 2^{jp} \mathbf{B}_j$$

whence

$$|X|_m = \left| \sum_{j=0}^{k-1} 2^{jp} \mathbf{B}_j \right|_m \tag{3.7}$$

$$= \left| \sum_{j=0}^{k-1} \left| 2^{jp} \mathbf{B}_j \right|_m \right|_m$$

EXAMPLE. Consider the computation of the residue of the 16-bit number 32015 with respect to the modulus 17. The binary representation of 32015 is 0111110100001111. We partition this into four 4-bit blocks—0111 1101 0000 1111—and then compute the residue as

$$|32015|_{17} = \left| \left| 7 \times 2^{12} \right|_{17} + \left| 13 \times 2^8 \right|_{17} + \left| 0 \times 2^4 \right|_{17} + \left| 15 \times 2^0 \right|_{17} \right|_{17}$$
$$= |10 + 13 + 0 + 15|_{17}$$
$$= 4.$$

END EXAMPLE

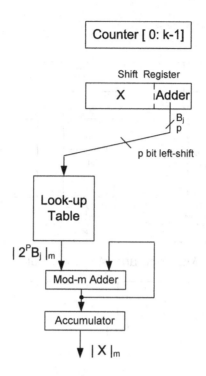

Figure 3.5: Sequential-parallel table-lookup converter

The simplest structure for the summation of Equation 3.7 is just a straightforward variant of that in Figure 3.3. Since that would give a purely sequential implementation, there is no gain in storing all the possible values of $\left|2^{jp}\mathbf{B}_j\right|_m$. The sets of values that the $\mathbf{B}_j$ blocks can take are the same set of 2^p values. Successive powers of two can be obtained by sequential multiplication. And multiplication by 2^p is just a p-bit hardwired left-shift. So we need store only the values of $\left|2^p\mathbf{B}_j\right|_m$. Thus we have the structure of Figure 3.5. The conversion will now take k cycles, and the amount of memory required is about $2^p \log_2 m$ bits.

The other extreme from the structure of Figure 3.5 is a highly-parallel arrangement in which all the possible values of $\left|2^{jp}\mathbf{B}_j\right|_m$ are stored in k lookup-tables. To compute the residue for a given number, k values are simultaneously read from all k tables and then added up in a tree of adders, a *multi-operand modular adder*. The basic structure is shown in Figure 3.6. Between the structure of Figure 3.5 and that of Figure 3.6, there exist several sequential-parallel ones; we leave it to the reader to investigate these.

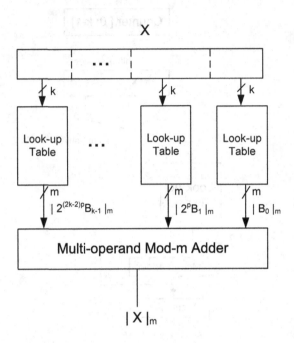

Figure 3.6: Parallel table-lookup converter

A straightforward way to implement a multi-operand modular adder as a tree of two-input combinational-logic modular adders Chapter 5. An alternative would be to use look-up tables to realize the modular adders; such a look-up table would take the two operands as addresses and return the result of the modular addition. The result would an all-table implementation, typically be implemented using ROMs. The choice between the two depends on the trade-off between performance (delay) and cost (area); power consumption may also be an issue. The ROMs will have a high regularity and density, which means low cost, but the performance will likely not match that of the combinational logic. Further discussion on the design of multi-operand modular adders will be found in [3].

The choice of block-size, p, above is seemingly arbitrary. So we may refer to it as "physical partitioning", in that depending on the technology used for realization, the best choice can be determined by design-space exploration. We next describe a technique that may be considered as a "logical partitioning", in that it is based on a property of residues. Both techniques may be combined; we shall leave details to the diligent reader.

Table 3.1: Periodicity properties of residues

Modulus m	Residue $2^n \bmod m$
3	1,2,1,2,1,...
5	1,2,4,3,1,2,4,3,..
6	1,2,4,2,4,2,4,...
7	1,2,4,1,2,4,...
9	1,2,4,8,7,5,1,2,4,8,7,...
10	1,2,4,8,6,2,4,8,6,...
11	1,2,4,8,5,10,9,7,3,6,1,2,4,5,8,10,9,7..
12	1,2,4,8,4,8,4,8,....
13	1,2,4,8,3,6,12,11,..
14	1,2,4,8,2,4,8,..
15	1,2,4,8,1,2,4,8,.....
17	1,2,4,8,16,15,13,9,..
18	1,2,4,8,16,14,10,2,4,8,...
19	,2,4,8,16,13,7,14,9,18,...
20	1,2,4,8,16,12,4,8,16,....
21	1,2,4,8,16,11,1,2,4,8,....

3.2.3 Sequential/parallel conversion: periodic partitioning

The *cyclic property* of $2^j \bmod m$ refers to the eventual repetition of residues as the value of j increases [4]. For example, with $m = 3$, the values of $2^j \bmod 3$, for $j = 0, 1, 2, \ldots$, repeat after the second residue: the residues are 1, 2, 1, 2, ... From the above (Equation 3.6), we know that the residues of X with respect to m are readily obtained if the residues $|2^j| m$ are available. But the approach outlined above is simplistic in that the basic idea is to store all of the latter residues. The approach outlined next relates the storage and the partitioning according to the inevitable repetition of the residues as j increases.

Table 3.1 lists the residues of powers of two with respect to several different moduli; no entry exists for m that is a power of two, as each residue will then be zero. We can observe from this table that some rows have only a few residues listed while other rows have longer lists of residues. The number of residues in each row depends upon the cyclic property of the residue with respect to that modulus. All residues are periodic, but the length of a period depends on the specific modulus. Thus, some residues have short period while others have a long period. Given a modulus, m, we shall refer to $m - 1$ as the *basic period* and to any shorter period as a *short period*.

Table 3.2: Period of repetition of residues

Odd m		Even m	
m	l	m	l
3	2	6	3
5	4	10	5
7	3	12	4
9	6	14	4
11	10	16	7
13	12	18	7
15	4	22	11
17	8	24	5
19	18	26	13
21	6	28	5

We can also observe from Table 3.1 other important properties of residues. For some moduli, the residues of 2^j (up to the point where the repetition starts) are distinct and therefore all $m - 1$ residues must be stored. As an example, consider the modulus 11. The residues $|2^j|_{11}$ are $1, 2, 4, 8, 5, 10, 9, 7, 3, 6$, for $j = 0, 1, 2, 3, 4, 5, 6, 7, 8, 9$. The residues do not repeat within the basic period determined by 11, but they do repeat from $j > 10$. Therefore, the maximum number of residues that must be stored is ten. For larger values of j, the same set of residues can be used because of the repetitive nature of residues beyond $j = 10$. Table 3.1 also shows that for some moduli, the residues repeat within the basic period determined by the modulus. For example, with the modulus 21, the residue-set $\{1, 2, 4, 8, 16, 11\}$ is obtained for $j = 0, 1, 2, 3, 4, 5$; for $j > 5$ the residues repeat. The periodicity of the residues is six, which is smaller than the basic period corresponding to 21. So in this case we need to store only six residues; for higher indices corresponding to the same modulus, the same set can be used; relative to the magnitude of the modulus, little memory is required in this case. The residue-set for modulus 6 is slightly different. The residue-set here is $\{1, 2, 4, 2, 4\}$, for $n = 0, 1, 2, 3, 4$. Observe that the residue-set is periodic after an initial period of unity: the residues repeat from 2, and this requires storing 1 in addition to 2 and 4.

Note that although Table 3.1 includes some special moduli (i.e. of the $\{2^n-1, 2^n, 2^n+1\}$ moduli-sets), conversion in these sets can be implemented quite easily using only combinational logic.

Table 3.2 lists the number, l, of residues that must be stored, for various moduli up to 21. (In practice large moduli are rarely used because of the imbalances they create in the implementation data-path.) As one would expect from the theoretical results of Chapters 1 and 2, l is large for prime m.

We now turn to the application of the cyclic nature of the residues to the implementation of forward conversion. Observe that the complexity in the computation of the residues of a given number depends on the periodicity of the residues and not on the number of bits in the number's representation. So we proceed as follows. If the period is t, the binary number whose residues are sought is partitioned into t-bit blocks. The sets of residues for the different t-bit blocks will be the same because of the periodicity: each block determines the same set of 2^t different values. The t-bit blocks are then added. The residue of any t-bit number and the result of the final modular additions, are obtained from a look-up table in which the residues of $2^j \bmod m$, $j = 0, 1, 2, \ldots, t$, are stored. Thus the basic hardware structures required are similar to those with arbitrary partitioning (Figures 3.6 and 3.7).

EXAMPLE. Consider the computation of the residue of the 16-bit number 32015 with respect to the modulus 17. The binary representation of 32015 is 0111110100001111. From Table 3.2, we find that the periodicity of the residues for 17 is 8. The 16-bit number is therefore partitioned into two 8-bit blocks that are then added:

$$01111101 + 00001111 = 10001100$$

The non-zero bits in the result correspond to $2^7, 2^3$, and 2^2. The residues of these, with respect to 17, are obtained from a lookup-table and then added in a multi-operand modular adder.

$$\begin{aligned} |32015|_{17} &= \bigl|\, |2^7|_{17} + |2^3|_{17} + |2^2|_{17} \,\bigr|_{17} \\ &= |9 + 8 + 4|_{17} \\ &= 4 \end{aligned}$$

END EXAMPLE

3.3 Arbitrary moduli-sets: combinational logic

Almost all forward-conversion architectures for arbitrary moduli-sets utilize a combination of ROM and combinational logic [11]. Nevertheless, it is possible to formulate forward conversion in terms of *modular exponentiation* and addition. This facilitates the design of converters that use only combinational logic. As in the preceding cases, the implementations can be one of several configurations—bit-serial, digit-serial, bit-parallel, and so forth —by a simple variation of the number of multiplexers used in the design. Since no table-lookup is required, it is possible to significantly reduce the complexity of the implementation by sharing some of the circuitry among forward converters for several moduli. We next explain the basic idea of modular exponentiation and then show its application to forward conversion.

3.3.1 *Modular exponentiation*

Modular exponentiation is based on the simple idea that the residue-set for any binary number can be obtained from the residues of powers-of-two representation of that binary number. Of course, this is essentially the same idea already used above; but we shall now see how a different variation on the theme can be used as a basis for the design a purely combinational-logic converter [1].

If X is represented in p bits, as $x_{p-1}x_{p-2}\cdots x_0$, then

$$X = \sum_{j=0}^{p-1} 2^j x_j$$

and

$$|X|_m = \left| \sum_{j=0}^{p-1} 2^j x_j \right|_m$$

$$= \left| \sum_{j=0}^{p-1} |2^j x_j|_m \right|_m$$

Now consider the residue representation of $|2^N|_m$, where N is an n-bit number, $s_{n-1}s_{n-2}\cdots s_0$, and m is the modulus of interest:

$$|2^N|_m = |2^{s_{n-1}s_{n-2}s_{n-3}\cdots s_0}|_m$$

By associating the s bits with their binary positional weights, this equation may be rewritten into

$$|2^N|_m = |2^{n-1}s_{n-1} \times 2^{n-2}s_{n-2} \times \cdots + 2^{n-k+1}s_{n-k+1}$$
$$\times 2^{n-k}s_{n-k} \times 2^{n-k-1}s_{n-k-1} \times \cdots \times 2^1 s_1 \times 2^0 s_0|_m$$

Since s_i is 0 or 1, we may modify the first k terms in this equation thus

$$|2^N|_m = \left|\left[(2^{2^{n-1}}-1)s_{n-1}+1\right]\left[(2^{2^{n-2}}-1)s_{n-2}+1\right]\cdots\right.$$

$$\cdots \left[(2^{2^{n-k+1}}-1)s_{n-k+1}+1\right]\cdots$$

$$\cdots \left.\left[(2^{2^{n-k}}-1)s_{n-k}+1\right] 2^{2^{n-k-1}s_{n-k-1}}\ldots 2^{2^1 s_1} 2^{s_0}\right|_m \quad (3.8)$$

Now let r_{n-i} be $\left|2^{2^{n-i}}-1\right|_m$, for $i = 1, 2, \ldots k$. Then substituting for the first k terms in Equation 3.8, we have

$$|2^N|_m = |\,|(r_{n-1}s_{n-1}+1)(r_{n-2}s_{n-2}+1)\cdots(r_{n-k+1}s_{n-k+1}+1)$$

$$(r_{n-k}s_{n-k}+1)|_m \, 2^{2^{n-k-1}s_{n-k-1}}\ldots 2^{2^1 s_1} 2^{s_0}\bigg|_m \quad (3.9)$$

which is of the form

$$|\,|(a_{n-1}+1)(a_{n-2}+1)\cdots(a_{n-k+1}+1)(a_{n-k}+1)|_m \cdots$$
$$\cdots 2^{2^{n-k-1}s_{n-k-1}}\ldots 2^{2^1 s_1} 2^{s_0}\bigg|_m$$

where $a_{n-i} + 1 = r_{n-i}s_{n-i} + 1$.
Expanding Equation 3.9, we have

$$|2^N|_m = \left|\left|\prod_{i=n-k}^{n-1} r_i s_i + \sum_{q=1}^{p_{k-1}} \prod_q C_{k-1}^k \cdots + \sum_{q=1}^{p_1} \prod_q C_1^k + 1\right|_m \right.$$

$$\left. \times 2^{2^{n-k-1}s_{n-k-1}} + \ldots 2^{2^1 s_1} 2^{s_0}\right|_m \quad (3.10)$$

where

$$p_{k-i} = C_{k-i}^k$$

$$= \frac{k!}{(k-i)!i!}$$

The term $\sum_{q=1}^{p_{k-i}} \prod_q C_{k-i}^k$ represents the sum of p_{k-i} products taken $(k-i)$ terms at a time, with each term being of the form $r_{n-i}s_{n-i}$. In Equation

3.10, the expression within the inner modular reduction on the right-hand side, is synthesized as a logic function g:

$$g(s_{n-1}, s_{s-2}, \ldots s_{n-k}) \triangleq \left| \prod_{i=n-k}^{n-1} r_i s_i + \sum_{q=1}^{p_{k-1}} \prod_q C_{k-i}^k \ldots + \sum_{q=1}^{p_1} \prod_{q=1} C_1^k + 1 \right|_m$$

The synthesis includes a modular reduction since this forms part of the function synthesized, which is done by giving alternating assigning 0 and 1 to each of the bits s_i. Equation 3.9 may then be rewritten by substituting the logic function for the product term:

$$\left| 2^N \right|_m = \left| g(s_{n-1}, s_{n-2}, \ldots s_{n-k}) 2^{2^{n-k-1} s_{n-k-1}} \ldots 2^{2^1 s_1} 2^{s_0} \right|_m$$

The bits $s_{n-k-1}, s_{n-k-2} \ldots, s_0$ appearing in the exponents are used to multiplex the function g. This requires that the function g be modified to obtain 2^{n-k} other logic functions, $g_j, j = 0, \ldots, 2^{n-k} - 1$. These functions are designed to take into account the positional weights associated with the bits $s_{n-k-1}, s_{n-k-1}, \ldots s_0$ that appear as exponents in the terms $2^{2^{n-k-1} s_{n-k-1}}$. In determining the logic functions g_j, simplifications involving the modular reductions are taken into account, thus eliminating the need for additional residue units when modular exponentiation is performed at lower levels. So

$$\left| g(s_{n-1}, \ldots, s_{n-k}) 2^{2^{n-k-1} s_{n-k-1}} \ldots 2^{2^1 s_1} 2^{s_0} \right|_m =$$

$$\begin{cases} g_0 (s_{n-1}, \ldots, s_{n-k}) s_{n-\bar{k}-1} \ldots, \bar{s}_0) \\ \text{or} \\ g_1 (s_{n-1} \ldots, s_{n-k}) s_{n-\bar{k}-1} \ldots \bar{s}_1 s_0) \\ \text{or} \\ \vdots \\ \text{or} \\ g_{n-1} (s_{n-1} \ldots s_{n-k}) s_{n-k-1} \ldots s_0) \end{cases} \quad (3.11)$$

In practical designs, the number of bits used to represent a modulus is usually quite small—very often just five or six bits—so the synthesis of the logic functions need not be very complex. The next example illustrates the procedure just described.

Detailed example

We now use a detailed example to illustrate modular-exponentiation forward -conversion. The forward translation consists of synthesizing the logic functions g_i to generate partial residues corresponding to each power of two that is required in the sum representing the given binary number. In computing partial residues, the bits that comprise the residues are multiplexed according to Equation 3.11.

Suppose we wish to evaluate $|8448|_{13}$. 8448 may be represented in powers-of-two form as $2^{13} + 2^8$. So

$$|8448|_{13} = ||2^{13}|_{13} + |2^8|_{13}|_{13}$$

and

$$\begin{aligned}|2^{s_3 s_2 s_1 s_0}|_{13} &= |2^{8s_3 + 4s_2 + 2s_1 + s_0}|_{13} \\ &= |(255 s_3 + 1)(15 s_2 + 1) 4^{s_1} 2^{s_0}|_{13} \\ &= |(3 s_3 s_2 + 8 s_3 + 2 s_2 + 1) 4^{s_1} 2^{s_0}|_{13} \qquad (3.12)\end{aligned}$$

Other choices are available in how to split $s_3 s_2 s_1 s_0$. Having made the choice above, we next look at the values that Equation 3.12 yields: for each of the four possibilities determined by $s_1 s_0$, we will determine a function gig, according to the values that $s_3 s_2$ can take.

When $s_1 s_0 = 00$, Equation 3.12 reduces to

$$|2^{s_3 s_2 s_1 00}|_{13} = |3 s_3 s_2 + 8 s_3 + 2 s_2 + 1|_{13}$$

For $s_3 s_2 = 00$, Equation 3.12 reduces to

$$\begin{aligned}|2^{0000}|_{13} &= |3 s_3 s_2 + 8 s_3 + 2 s_2 + 1|_{13} = 1 \\ &= 1\end{aligned}$$

For $s_3 s_2 = 01$, $s_3 s_2 = 10$ and $s_3 s_2 = 11$, Equation 3.14 gives the results 3, 9, and 1 respectively. The various values are shown in Table 3.4. Minimization produces the bit-sequence $(s_3 \bar{s}_2, 0, \bar{s}_3 s_2, 1)$, which corresponds to the weights 8,4,2,1. That is, the function $g_0 = s_3 \bar{s}_2 + 2 \bar{s}_3 s_2 + 1$.

Table 3.4: Truth table for g_0

$s_3 s_2$	8	4	2	1
00	0	0	0	1
01	0	0	1	1
10	1	0	0	1
11	0	0	0	1

Table 3.5: Truth table for g_1

s_3s_2	8	4	2	1
00	0	0	1	0
01	0	1	1	0
10	0	1	0	1
11	0	0	1	0

When $s_1s_0 = 01$ Equation 3.12 reduces to

$$|2^{s_3s_201}|_{13} = |(3s_3s_2 + 8s_3 + 2s_2 + 1) \times 2|_{13}$$
$$= |6s_3s_2 + 3s_3 + 4s_2 + 2|_{13}$$

The four different combinations of s_3s_2 are shown in Table 3.5. Minimization produces the logic function $g_1 = (0, \bar{s}_3s_2 + s_3\bar{s}_2, \bar{s}_3 + s_2, s_3\bar{s}_2)$.

Lastly, Tables 3.6 and 3.7 show the mappings for Equation 3.12 when $s_1s_0 = 10$ and $s_1s_0 = 11$ respectively:

$$|2^{s_3s_210}|_{13} = |(3s_3s_2 + 8s_3 + 2s_2 + 1) \times 4|_{13}$$
$$= |12s_3s_2 + 6s_3 + 8s_2 + 4|_{13}$$

and

$$|2^{s_3s_211}|_{13} = |(3s_3s_2 + 8s_3 + 2s_2 + 1) \times 8|_{13}$$
$$= |11s_3s_2 + 12s_3 + 3s_2 + 8|_{13}$$

Table 3.6: Truth table for g_2

s_3s_2	8	4	2	1
00	0	1	0	0
01	1	1	0	0
11	0	1	0	0
10	1	0	1	0

Table 3.7: Truth table for g_3

$s_3 s_2$	8	4	2	1
00	1	0	0	0
01	1	0	1	1
11	1	0	0	0
10	0	1	1	1

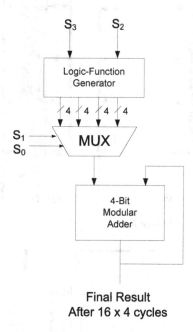

Figure 3.7: Sequential implementation of the forward converter

Minimizing Tables 3.6 and 3.7 yields the functions g_2 and g_3 as $(\bar{s}_3 s_2 + s_3 \bar{s}_2, \bar{s}_3 + s_2, s_3 \bar{s}_2, 0)$ and $(\bar{s}_3 + s_2, s_3 \bar{s}_2, \bar{s}_3 s_2 + s_3 \bar{s}_2, \bar{s}_3 s_2 + s_3 \bar{s}_2)$, respectively. The conversion hardware consists of sixteen columns of four 1-bit multiplexers each. Strictly, four multiplexers are sufficient as there can at most be four bits in a residue in this case. Since only s_1 and s_0 are used for multiplexing-control, two to four multiplexers are needed. The most significant bits of all four g functions are applied to the first multiplexer and the least significant bits are applied to the last multiplexer.

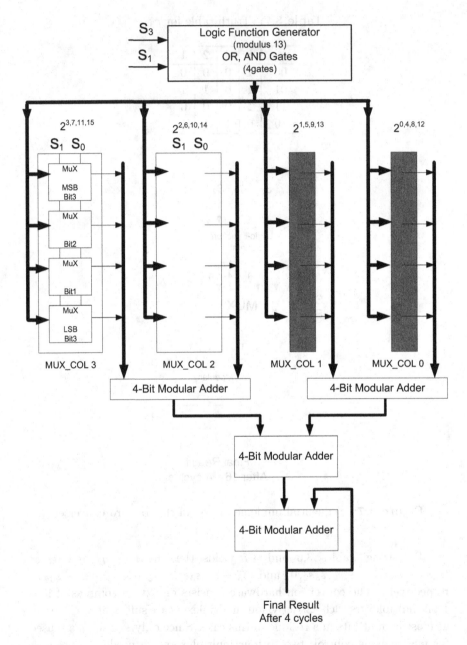

Figure 3.8: Sequential-parallel implementation of the forward converter

Consider, for example, 8448 mod 13. In terms of powers of two, we have $|8448|_{13} = \big| |2^{13}|_{13} + |2^8|_{13} \big|_{13}$. Here $N = 13, s_3s_2s_1s_0 = 1101$, and $s_1s_0 = 01$ are used as multiplexer-inputs and for the function g_1. Using four one-bit multiplexers such that the most significant bits of g_0 through g_3 are inputs to the first multiplexer and the least significant bits are inputs to the fourth multiplexer, the output is obtained in one clock cycle. The output corresponding to g_1 is $\{0, (1 \oplus 1), (\bar{1}\ or\ 1), 0\}$; so $g_1 = 0010 = 2$. Similarly, for $N = 8$, g_0 is the output, and the residue corresponding to this is $1001 = 9$. So $|8448|_{13} = |2 + 9|_{13} = 11$.

In our specific example here, 2^{13} and 2^8 are the only non-zero powers of two present. The corresponding binary equivalents of 13 and 8 are 1101 and 1000. Two modular exponentiations are performed with s_1s_0 as 01 and 00, respectively, and the logic functions g_1 and g_0 are multiplexed. Functions g_1 and g_0 when evaluated for $s_3s_2 = 11$ and $s_3s_2 = 10$ yield $g_1 = (0, 0, 1, 0) = 2$ and $g_0 = (1, 0, 0, 1) = 9$, as can be seen from Tables 3.4 and 3.5, respectively. These results are then added in modular adders, and the final residue is 11. Figures 3.7 and 3.8 show the sequential and sequential-parallel implementations of the converter. The sequential-parallel design uses four 4-bit, 4-to-1 multiplexers and requires 4 clock cycles to complete the conversion. The sequential design uses only one 4-to-1 multiplexer but requires 16 × 4 clock cycles, since there are four bits in each residue and 16-bit numbers are being considered.

Architectural realization

We have shown that in the forward conversion of a binary number to its RNS equivalent, modular exponentiation is performed on each non-zero bit of the binary number. For the corresponding hardware implementation, there are three alternatives that exist for carrying out this translation: sequential, parallel, or a combination of both parallel and sequential methods. In this section, we shall briefly describe these three methods. This discussion will include cost, complexity issues, and performance factors associated with the conversion. Since a dynamic range of thirty-two bits has been found to be sufficient for most digital-signal-processing applications, we shall assume the moduli-set $\{3, 5, 7, 11, 13, 17, 19, 23, 29, 31\}$. So, five bits will suffice to represent a modulus. The logic circuits that are used to combine and generate the terms that make up g_i can be shared among different moduli, which will yield significant savings if the numbers to be converted are fairly large.

In sequential conversion, we perform modular exponentiation on each non-zero bit. Therefore, in the conversion of a 32-bit number, the exponentiation is repeated thirty-two times. We have a choice between using a single 1-bit multiplexer or using five 1-bit multiplexers (i.e. one 5-bit multiplexer). In using a single 1-bit multiplexer, every bit in a 5-bit residue is multiplexed; so the circuit requires five clock cycles for each modular exponentiation. Since, there are thirty-two bits in the number to be converted, the total number of clock cycles required is 160 (i.e. 32 × 5). Evidently, the operational delay can be greatly reduced—by a factor of five—if five 1-bit multiplexers are used instead, as all five bits in a residue can then be multiplexed at the same time. We also have a choice in the number of bits used for multiplexer-control: we may use two bits for multiplexer-control and leave the other three bits for logic synthesis, or use three bits for multiplexer-control and the remaining two bits for logic synthesis, and so forth. Using fewer bits for multiplexer-control greatly reduces the circuit complexity. For the 32-bit case under consideration, we have determined that using two bits for multiplexer-control resulted in the least costs (as measured by the number of gates) in the implementation of the converter. The partial residues obtained for each modular exponentiation are added in a multi-operand modular adder.

In the parallel computation of residues, modular exponentiation is carried out simultaneously on all bits of a given binary number. This requires the use of thirty-two 1-bit multiplexers for a 32-bit number and five clock cycles for 5-bit residues. The speed of residue computation can be further increased by using five times as many multiplexers, as this enables all modular exponentiations to be performed in one clock cycle; partial residues are then added using a tree of adders (a multi-operand modular adder). This method has one drawback: if the converter is realized in current VLSI technology, the routing complexity is likely to be high and will be directly proportional to the number of bits in a residue.

Figure 3.9 shows an architecture for the parallel implementation of the conversion algorithm. The combinational-logic circuit required to generate the bits in the sequences is common to all the multiplexers. In fact, the same circuit can be shared among converters for several moduli. The conversion requires only one clock cycle because of the high parallelism inherent in the implementation.

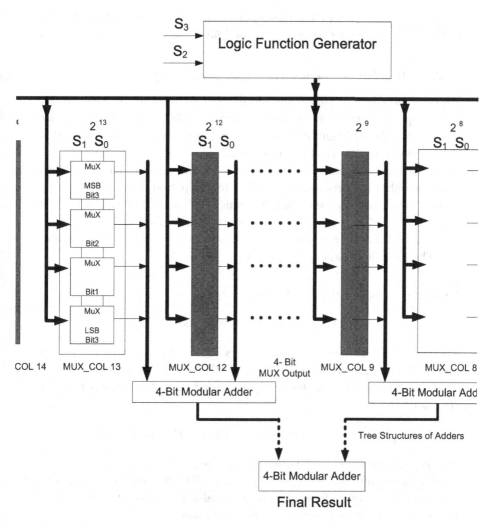

Figure 3.9: Parallel modular-exponentiation forward converter

An alternative to either the purely sequential or purely parallel method is a combination of a bit of both. In this case, speed is compromised to somewhere between that of the sequential method and that of the parallel method, but the hardware costs increase to just a little more than that of the sequential approach. The sequential-parallel method is implemented by partitioning the given binary number into several blocks and then concurrently carrying out modular exponentiation on all the partitions. In this case, the the modular exponentiation and final additions can be pipelined.

3.3.2 Modular exponentiation with periodicity

We next show how modular exponentiation can be combined with periodicity property of residues to design a forward converter of low hardware complexity. We shall initially assume that the moduli are odd, as is often the case in many practical implementations.

In applying the periodicity properties to compute residues, the given binary number is partitioned according to the period, basic or short, associated with the corresponding modulus. Let p be that period. Then 2^n, 2^{n+p}, 2^{n+2p},... will all have the same residues. The partitioned blocks are then added, with the sum reduced appropriately relative to the modulus. The following example illustrates the procedure.

EXAMPLE. Consider the 32-bit number

00011001 00011010 11000110 00111100

The decimal equivalent of this number is 421185084, and its residue with respect to 23 is 22. From Table 3.2, the periodicity of the residues is found to be 11. The number is therefore partitioned into 11-bit blocks that are then added:

00001100100 + 01101011000 + 11000111100 = 100111111000

Forward conversion is then performed on the result, using modular exponentiation on only eleven bits, as opposed to all thirty-two bits as would be the case with a straightforward combinational-logic converter. This reduces the number of adders required for the conversion.
END EXAMPLE

In many applications, the moduli are odd, except for those of the form 2^n, which occur in the special moduli-sets; and reduction modulo 2^n requires no hardware. Nevertheless, there are cases in which even moduli other than 2^n do occur; for example, in the moduli-set $\{2m-1, 2m, 2m+1\}$. So, contrary to what might initially appear to be the case, there is also a need for forward conversion in the case of even moduli. Residue computation in this case is performed in a slightly different manner: Even moduli are represented as a product of powers-of-two and an odd number. The following property is then applied in the evaluation of residues. The residue of an integer X modulo m can be expressed as

$$|X|_m = \left|\left\lfloor \frac{X}{2^p} \right\rfloor\right|_y 2^p + |X|_{2^p} \qquad (3.13)$$

where m, the even modulus, is expressed as $2^p y$, with y an odd number. $|X|_{2^p}$ is just the least significant p bits of X. If X is represented in k bits, then the first term in Equation 3.13 is just the $k - p$ most significant bits of the X. The evaluation of $|\lfloor X/2^p \rfloor|_y$ is as above, for odd moduli, since y is odd. The proof of Equation 3.13 is as follows.

$$X = QP + R$$

where Q is the quotient obtained by dividing X by P and R is the residue. Suppose P is even, composite, and not in the 2^n set. Let P denote the value $y2^p$. Then

$$|X|_P = R$$

Now, X may also be expressed as

$$X = Q_1 2^p + R_1$$

where Q_1 is the quotient and R_1 is the residue obtained from dividing X by 2^p. So $0 \le R_1 < 2^p$, and $R_1 = |X|_{2^p}$. And Q_1 may be expressed as

$$Q_1 = Q_2 y + R_2$$

where Q_2 is the quotient and R_2 is the remainder obtained by dividing Q_1 by y. So $0 \le R_2 < y$, and $Q_1 = \lfloor X/2^p \rfloor$ from Equation 3.15. Since R_2 is the residue of Q_1,

$$R_2 = |Q_1|_y = \left|\left\lfloor \frac{X}{2^p} \right\rfloor\right|_y$$

Therefore

$$X = QP + R$$
$$= Q_1 2^p + R_1$$

From the preceding

$$QP + R = (Q_2 y + R_2) 2^p + |X|_{2^p}$$

$$= Q_2 y 2^p + R_2 2^p + |X|_{2^p}$$

$$= Q_2 P + \left|\left\lfloor \frac{X}{2^p} \right\rfloor\right|_y 2^p + |X|_{2^p} \qquad (3.14)$$

Comparing $X = QP + R$ with Equation 3.14, we deduce that

$$R = \left|\left\lfloor \frac{X}{2^p} \right\rfloor\right|_y 2^p + |X|_{2^p}$$

The following example illustrates the procedure.

EXAMPLE. Let us determine the residue, with respect to 24, of the same 32-bit number as in the last example:

00011001 00011010 11000110 00111100

The decimal equivalent of this number is 421185084, and its residue with respect to 24 is 12. The modulus is expressed as a composite number: $24 = 2^3 \times 3$. The partial residue $|X|_{2^3}$ is simply the three least significant bits of X, i.e. 100. The remaining 29 bits are partitioned into 2-bit blocks since the periodicity of 3 is 2. The 2-bit blocks are then added modulo 3. The modular exponentiation is then performed and the partial residue obtained is to be 1. Combining the two partial residues, we get 1100, which in decimal is 12. END EXAMPLE

3.4 Summary

As is the case with all residue operations, implementations for forward conversion may be classified according to the ease of design and realization; that is, according to whether the moduli are of the special type, or whether they are not. With the former moduli, the converters are almost always of pure combinational-logic; and with the latter most involve the use of some table-lookup, although modular exponentiation facilitates the use of just combinational logic. The best converters for the arbitrary moduli invariably exploit particular properties of residue, the most significant of which is the fact that residues with respect to a given modulus repeat after some point. Also, the various techniques described can be fruitfully combined to take advantage of their different merits.

References

(1) B. Premkumar. 2002. A formal framework for conversion from binary to residue numbers. *IEEE Transactions on Circuits and System II*, 46(2):135–144.

(2) N. S. Szabo and R. I. Tanaka. 1967. *Residue Arithmetic and its Applications to Computer Technology*, McGraw-Hill, New York.
(3) S. Piestrak. 1994. Design of residue generators and multioperand modular adders using carry save adders. *IEEE Transactions on Computers*, 42(1):68–77.
(4) A. Mohan. 1999. Efficient design of binary to RNS converters. *Journal of Circuits and Systems*, 9(3/4): 145–154.
(5) A. Mohan. 2002. *Residue Number Systems: Algorithms and Architectures*. Kluwer Academic Publishers, Dordrecht.
(6) B. Parhami and C. Y. Hung. 1994. "Optimal table lookup schemes for VLSI implementation of input/output conversions and other residue number operations". In: *VLSI Signal Processing VII*, IEEE Press, New York.
(7) F. Barsi. 1991. Mod m arithmetic in binary systems. *Information Processing Letters*, 40:303–309.
(8) D. K. Banerji and J. A. Brzozowski. 1972. On translation algorithms in RNS. *IEEE Transaction on Computers*, C-21:1281–1285.
(9) G. Alia and E. Martinelli. 1984. A VLSI Algorithm for direct and reverse conversion from weighted binary number to residue number system. *IEEE Transaction on Circuits and Systems*, 31(12):1425–1431.
(10) G. Alia and E. Martinelli. 1990. VLSI binary-residue converters for pipelined processing. *The Computer Journal*, 33(5):473–475.
(11) R. M. Capocelli and R. Giancarlo. 1998. Efficient VLSI networks for converting an integer from binary system to residue number system and vice versa. *IEEE Transactions on Circuits and System*, 35(11):1425–1431.
(12) G. Bi and E. V. Jones. 1988. Fast Conversion betweeen binary and residue numbers. *Electronic Letters*, 24(9):1195–1997.
(13) B. Vinnakota and V. B. B. Rao. 1994. Fast conversion techniques for binary-residue number systems. *IEEE Transaction on Circuits and Systems*, 14(12):927–929.

Chapter 4

Addition

The main topics of this chapter are algorithms for addition in residue number systems and the hardware implementations of these algorithms. By implication, this also includes subtraction, which is usually implemented as the addition of the negation of the subtrahend. Hardware implementations for residue arithmetic may be based on look-up tables (realized as ROMs), or pure combinational logic, or a combination of the two. The combinational-logic units in such cases are exactly those for conventional arithmetic or are derived from those, whence our remarks in Chapter 1 on the relationships between residue notations, on the one hand, and arithmetic and conventional notations and arithmetic, on the other. We shall therefore start with a review of the major types of conventional adders and then show how these may be used as the basis for residue adders. The reader who is already familiar with this material may quickly skim the first section of the chapter, primarily for notation and terminology.

The second section of the chapter deals with addition relative to an arbitrary modulus. Such addition cannot be implemented as efficiently as conventional addition, even though the same underlying units may be used in both cases. The third and fourth sections are on addition relative to two special moduli, $2^n - 1$ and $2^n + 1$. The importance of these moduli arises from the fact that the moduli-set $\{2^n - 1, 2^n, 2^n + 1\}$ is an especially popular one, as are its extensions, largely because of the relative ease with which arithmetic operations with respect to these moduli can be implemented. Addition modulo 2^n is just conventional unsigned addition, and addition modulo $2^n - 1$ is very similar to 1s-complement addition. Addition modulo $2^n + 1$ is, however, less straightforward.

In discussing basic arithmetic operations in conventional number systems, it would be natural to include some remarks on overflow. We shall not do so here, because determining overflow in a residue number system is a rather difficult operation and is one that is best included with other operations of the same fundamental nature (Chapter 6).

We conclude these introductory remarks with two notes on terminology. First, because, digit-wise, the addition of two numbers represented in residue notation can be performed concurrently, and independently, by adding all corresponding digits, the following discussions on residue addition are limited to just operations on digit-pairs. That is, we shall be discussing the design of digit-slices. Second, unqualified uses of the term "adder" will be references to conventional adders; for residue arithmetic, we shall use "residue adder", sometimes with an indication of the modulus in use.

4.1 Conventional adders

There are many basic designs for adders, and many more designs can be obtained as "hybrids" of these basic designs. In what follows we shall briefly describe the major basic designs. The choice of the particular adders covered has been determined by both the frequency of their usage in conventional arithmetic and ease with which they can be applied to residue arithmetic: certain types of adders, e.g. the conditional-sum and parallel-prefix adders, are particularly important because they can easily be adapted for addition with special moduli (such as $2^n - 1$ and $2^n + 1$). For more comprehensive discussions of adders, the reader is referred to [12, 13, 21].

The designs covered here are those of the carry-ripple, carry-skip, carry-lookahead, carry-select, conditional-sum, and parallel-prefix adders. The ripple adder is very cheap and effective for short word-lengths and may therefore be useful in the many applications of residue arithmetic in which small moduli are used. (The basic technique can also be combined with the use of other techniques in order to implement large adders.) For word-lengths for which the ripple adder may be too slow, the carry-skip adder is one option for better performance at relatively low cost.[1] All of the other adder designs described here are "high-performance" but entail much higher costs: The carry-lookahead adder is theoretically the fastest design, but in its pure form it is practical only for short word-lengths. This is usually a

---

[1] For ASICs, a good measure of cost is chip area, of which gate-count and the number of interconnections comprise a reasonable approximation. For FPGAs, the number of configurable-logic blocks, or something similar, is a good measure.

problem with conventional addition, but is not necessarily the case with typical residue arithmetic, given the many applications for which small moduli suffice. Carry-select addition can be used with other underlying basic techniques and is also "naturally" suited to certain cases in residue addition—those in which two intermediate results are generated and one is then selected to be the final result. The conditional-sum adder may be viewed as a particular instance of combining carry-lookahead and carry-selection principles; it is suitable for a wide range of word-lengths. For the realization of fast addition in current technology, the parallel-prefix adder has the best performance-cost characteristics and, essentially, allows carry-lookahead principles to be practically employed for a wide range of word-lengths.

4.1.1 Ripple adder

Excluding a *serial adder*, which consists of just a full adder and 1-bit storage, the simplest possible adder is the *carry-ripple* (or just *ripple*) adder. Such an adder consists of a series of full adders in a chain, with the carry output from one full adder connected to the carry input of the next full adder. The design of an n-bit ripple adder that takes two operands, $A \triangleq A_{n-1}A_{n-2}\ldots A_0$ and $B \triangleq B_{n-1}B_{n-2}\ldots B_0$, and produces a sum, $S \triangleq A_{n-1}S_{n-2}\ldots S_0$ is shown in Figure 4.1. C_{-1} is the carry into the adder and is usually 0 for unsigned arithmetic; C_{n-1} is the carry out of the adder. The logic equations for a full adder are

$$S_i = (A_i \oplus B_i) \oplus C_{i-1} \quad (4.1)$$
$$C_i = A_iB_i + (A_i \oplus B_i)C_{i-1} \quad (4.2)$$

and a corresponding logic diagram is shown in Figure 4.2. Throughout the text we shall assume that this is the implementation for a full adder, although other variations are possible

If performance is measured in terms of logical date-delays, then the serial adder appears to be rather slow, because the full adders cannot always operate in parallel. In general, the full adder at stage i has to wait for a possible carry from stage $i-1$, which in turn has to wait for a possible carry from stage $i-2$, and so forth. The operational time is therefore $O(n)$, in contrast with the $O(\log n)$ of the fastest adders. Nevertheless in current technology it is possible to construct realizations of the ripple adder whose operational times are quite small. The basic technique in such realizations is a very fast carry-path, commonly known as a *Manchester carry-chain*; the resulting adder is a *Manchester adder*.

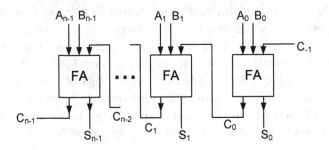

Figure 4.1: Ripple adder

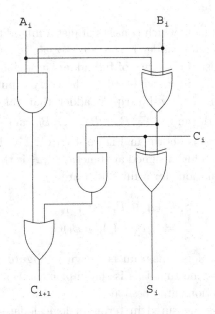

Figure 4.2: Full adder

The Manchester adder may be viewed as a carry-ripple adder in which the carry-path consists of switches instead of the usual gates, with the switches being considerably faster than the gates. Figure 4.3 depicts a single bit-stage in a Manchester carry-chain. It consists of three switches[2], S_G, S_P, and S_K, that operate in mutual exclusion. When both operand bits are 1, the switch S_G closes and the carry-out of the stage is 1, whatever the carry-in is. When one of the operand bits is 0, the other is 1, and carry-in is 1, the switch S_P closes and the carry-out of the stage is 1. And when both operand bits are 0, the switch S_K closes and the carry-out of the stage is 0, whatever the carry-in is. Figure 4.4 shows one possible CMOS realization of Figure 4.3; many other variants are possible. This is evidently a more efficient structure than would be directly obtained from the corresponding gates of Figure 4.3.

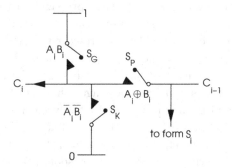

Figure 4.3: Stage of Manchester carry-chain

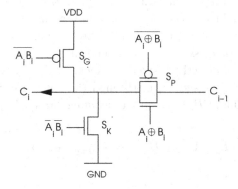

Figure 4.4: CMOS realization of Manchester carry-chain stage

---

[2]The bit-pair of a stage are such that it can *generate* a carry, or *propagate* an incoming carry, or *kill* an incoming carry.

For small word-lengths, Manchester adders are very fast and cost very little. Indeed, for such word-lengths the ripple adder is competitive with the more complex designs that logically have much smaller operational times (as measured by time-complexity). This is significant, given that many applications of residue arithmetic involve small moduli. The Manchester adder is, when combined with other techniques, also useful as a building block in large adders.

4.1.2 Carry-skip adder

The *carry-skip adder* (also known as a *carry-bypass adder*) is, essentially, the result of improving the (logical) carry-propagation times in the basic ripple adder, by having carries skip stages that they would definitely ripple through. Suppose the two numbers represented by the binary patterns $A_{12}A_{11}A_{10}A_9 010101 A_2 A_1 A_0$ and $B_{12}B_{11}B_{10}B_9 101010 B_2 B_1 B_0$ are added, any carry from the addition of bits 0 through 2 will ripple through the middle 0s and 1s stages and into the stages that add bits 9 through 12. Now, the patterns of 0s and 1s is such that any incoming carry from the addition of A_2 and B_2 always ripples through that middle portion and into the last four stages. Therefore, the addition in the latter stages may be started immediately, with a carry-in of 1, as soon as there is a carry from bit 2, while that carry is also rippled through, as usual, to form the sum bits in positions 3 through 8. So one may view the carry C_8 as being (a copy of) the carry C_2 that has skipped stages 3 through 8.

A carry-skip adder is therefore essentially a ripple adder (typically a Manchester adder) that has been divided into several *blocks*, each of which is equipped with *carry-skip* logic that determines when the carry into the block can be passed directly to the next block. The carry-skip logic is quite simple: A carry into stage i goes through the stage if either $A_i = 1$ and $B_i = 0$ or $A_i = 0$ and $B_i = 1$. So an incoming carry, C_{j-1}, skips a block, j, of size m if

$$[(A_j \oplus B_j)(A_{j+1} \oplus B_{j+1}) \cdots (A_{j+m-1} \oplus B_{j+m-1})]C_{j-1} \quad (4.3)$$
$$\triangleq (P_j P_{j+1} \ldots P_{j+m-1})C_{j-1}$$

(P_i is called the *carry-propagate* signal for stage i.) Therefore, the carry into block $j+1$ is either the carry, C_{j-1}, that skips block j, or is the carry, $\widetilde{C}_{j+m-1}$, that comes out the last stage of block j, having been generated somewhere in the block. The carry-skip logic for a block therefore nominally consists of one AND gate for each P signal, one AND gate to combine the

P signals, and an OR gate to select either source of carry; these implement the equation for the carry into the next block:

$$C_{j+m-1} = P_j P_{j+1} \ldots P_{j+m-1} C_{j-1} + \widetilde{C}_{j+m-1}$$
$$\stackrel{\triangle}{=} P_j^{j+m-1} C_{j-1} + \widetilde{C}_{j+m-1}$$

(P_j^{j+m-1} is called the *block carry-propagate* signal for block j.) Note that the P_i signals are already required to form the sum bits and therefore do not necessitate any extra logic.

It is common in realizations to replace the nominal OR gate with a multiplexer, so that the carry-skip logic then corresponds to the equation

$$C_{j+m-1} = P_j^{j+m-1} C_{j-1} + \overline{P}_j^{j+m-1} \widetilde{C}_{j+m-1}$$

Figure 4.5 shows a 16-bit adder, of 4-bit blocks, designed on this basis. The signal T_i is similar to P_i and is explained in Section 4.1.3.

What the carry-skip logic does is break up potentially long carry-propagation paths through the full adders. Assuming blocks of the same size, the longest chain will be one that starts in the second stage of the first block, ripples through to the end of that block, runs along the carry-skip path (of one OR gate and one AND gate per stage), and finally enters the last block and ripples through to the penultimate stage of that block. Speeding up such an adder requires two seemingly contradictory approaches. On the one hand, reducing the inter-block rippling requires that blocks be small. But on the other hand, reducing the length of the carry-skip path requires that the number of blocks be low, and, therefore, that the blocks be large. The obvious compromise is to have blocks of different sizes, arranged in such a way that the block sizes increase from either end of the adder. For example, for a 16-bit adder, one good partitioning is the set of block-sizes [1, 2, 3, 4, 3, 2, 1]. It should, however, be noted that for a given adder, the block-sizes depend on the implementation of the adder, as well as its realization in a given technology. The important factor is the ratio between the time required for a carry to ripple through a bit-slice and the time required to skip a block. The one-bit difference in the 16-bit example above is optimal for a 1:1 ratio, but clearly other ratios (integer and non-integer) are possible. Further discussions of carry-skip adders will be found in [7, 8].

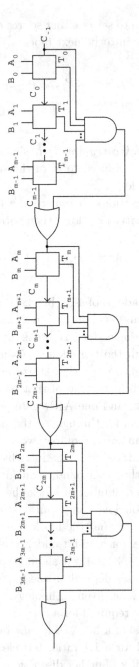

Figure 4.5: Carry-skip adder

The operational time for an optimally partitioned[3] n-bit carry-skip adder is $O(\sqrt{n})$. For large values of n, this is well above the $O(\log n)$ operational times of the fastest adders. When n is large better performance can be achieved by again applying the same basic technique: the original carry-skip chain is itself partitioned to yield another, second-level, carry-skip chain. In theory, this can be extended arbitrarily, but two levels of carry-skip logic appears to be the current practical limit. The operational time for an L-level carry-skip adder is $O(\sqrt[L]{n})$.

4.1.3 Carry-lookahead adders

Carry-lookahead is arguably the most important technique in the design of fast adders, especially large ones. In straightforward addition, e.g. in a ripple adder, the operational time is limited by the (worst-case) time allowed for the propagation of carries and is proportional to the number of bits added. So faster adders can be obtained by devising a way to determine carries before they are required to form the sum bits. Carry-lookahead does just this, and, in certain cases the resulting adders have an operational time that is independent of the operands' word-length.

A carry, C_i, is produced at bit-stage i if either one is *generated* at that stage or if one is *propagated* from the preceding stage. So a carry is generated if both operand bits are 1, and an incoming carry is propagated if one of the operand bits is 1 and the other is 0. Let P_i and G_i denote the generation and propagation, respectively, of a carry at stage i, A_i and B_i denote the two operands bits at that stage, and C_{i-1} denote the carry into the stage. Then we have

$$G_i = A_i B_i \tag{4.4}$$

$$P_i = A_i \oplus B_i \tag{4.5}$$

$$C_i = G_i + P_i C_{i-1} \tag{4.6}$$

The last of these is a reformulation of Equation 4.2, and Equation 4.1 may now be rewritten as $S_i = P_i \oplus C_{i-1}$, which allows the use of shared logic to produce S_i and P_i. In some cases it is convenient to replace the propagation function[4], P_i, by the *transfer function*, T_i, which combines

---

[3]For uniform-size blocks, the operational time is $O(n)$, but with smaller constants than for the ripple adder.

[4]In current CMOS technology, for example, XOR gates have a larger operational delay than AND/OR/NAND/NOR gates

both propagation and generation, since

$$T_i = G_i + P_i$$
$$= A_i + B_i$$

P_i would still be needed to form the sum. and can be computed concurrently with other signals, but if only T_i is formed initially. Then P_i may be obtained as

$$P_i = \overline{G_i}\,\overline{T_i}$$

Another function that is frequently used, instead of P_i, to propagate carries is the *kill* function, K_i. This function is implicit in the Manchester carry-chain and is defined as

$$K_i = \overline{A_i}\,\overline{B_i}$$

Evidently, $K_i = \overline{T_i}$ and $P_i = \overline{G_i + K_i}$. There are in fact several other functions that may be used to propagate carries [4], but we shall stick to just P_i.

If we unwind the recurrence (Equation 4.6) for C_i, for different values of i, then we end up with

$$C_0 = G_0 + P_0 C_{-1}$$
$$C_1 = G_1 + P_1 G_0 + P_1 P_0 C_{-1}$$
$$C_2 = G_2 + P_2 G_1 + P_2 P_1 G_0 + P_2 P_1 P_0 C_{-1}$$
$$\vdots$$

$$C_i = G_i + P_i G_{i-1} + P_i P_{i-1} G_{i-2} + \cdots + P_i P_{i-1} P_{i-2} \cdots P_0 C_{-1} \quad (4.7)$$

where C_{-1} is the carry into the adder.

Equation 4.7 for C_i states that there is a carry from stage i if there is a carry generated at stage i, or if there is a carry that is generated at stage $i-1$ and propagated through stage i, or if ..., or if the initial carry-in, C_{-1}, is propagated through stages $0, 1, \ldots, i$. The complete set of equations show that, in theory at least, all the carries can be determined independently, in parallel, and in a time (three gate delays) that is independent of the number of bits to be added. The same is also therefore true for all the sum bits, which require only one additional gate delay. Figure 4.6 shows the complete logical design of four-bit carry-lookahead adder.

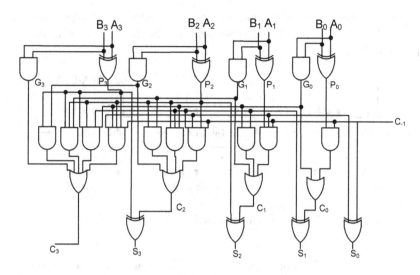

Figure 4.6: 4-bit carry-lookahead adder

Compared with a ripple adder, as well as some of the other adders described above, a pure carry-lookahead adder has high logic costs. Furthermore, high fan-in and fan-out requirements can be problematic: the fan-out required of the G_i and P_i signals grows rapidly with n, as does the fan-in required to form C_i. For sufficiently large values of n, the high fan-in and fan-out requirements will result in low performance, high cost, or designs that simply cannot be realized. In contrast with adders for conventional arithmetic, however, the moduli used in many applications of residue arithmetic are quite small (i.e. n is small), and carry-lookahead adders of the type depicted in Figure 4.6 may not be impractical. For large values of n, there are "hybrid" designs that combine carry-lookahead with other techniques, such as carry-rippling; nevertheless all such designs result in adders that are not as fast as pure carry-lookahead adders. Even when practical, the irregular structure and lengthy interconnections, exhibited in Figure 4.6 , are not ideal for VLSI implementation.

One straightforward way of dealing with the fan-in and fan-out difficulties inherent in the pure carry-lookahead adder is to split an n-bit adder into N (where $1 < N < n$) blocks of m bits each, arranged in such a way that carries within blocks are generated by carry-lookahead but carries between blocks are rippled. Such an arrangement is shown in Figure 4.7 , where CLA represents a complete lookahead adder of the type shown in Figure 4.6. We shall refer to this adder as the *Ripple Carry-Lookahead Adder* (RCLA).

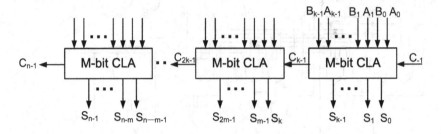

Figure 4.7: Ripple carry-lookahead adder (RCLA)

A different design for a practical carry-lookahead adder can be obtained by, essentially, reversing the basic design principle of the RCLA: carries within blocks are now rippled, but those between blocks are produced by lookahead. The implicit requirement here is that n be large enough and m small enough that the gain in speed from using lookahead outweighs the loss from rippling. Figure 4.8 shows the organization of such an adder, which we shall refer to as the *Block Carry-Lookahead Adder* or just *BCLA*. The *carry-lookahead* (CL) unit implements the lookahead equation for a single block-carry and has the logical details shown in Figure 4.9.

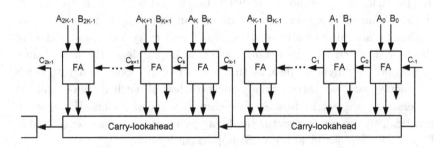

Figure 4.8: Block carry-lookahead adder (BCLA)

For large values of n, the performance of the RCLA will be limited by the inter-block rippling time, and that of the BCLA will be limited by the rippling between the block-lookahead units. Fan-in and fan-out will also be problematic with the BCLA. A "natural" way to get around these difficulties is to again apply the same principles; that is, to use an additional level of lookahead. There are many variations on that design-theme, but we shall here consider just one; for others, the reader is referred to [12].

To obtain two-level BCLA, an n-bit adder is divided into N m-bit blocks that are then grouped into M-block *superblocks*. The first level of lookahead is largely as before, and the second level of lookahead is on the superblocks. At the superblock level, the carry signals are produced through *auxiliary* propagation and generation signals, P_i^j and G_i^j, which are analogous to the P and G signals:

$$P_i^j = P_j P_{j-1} \cdots P_i \tag{4.8}$$
$$G_i^j = G_j + P_j G_{j-1} + P_j P_{j-1} G_{j-2} + \cdots + P_j P_{j-1} \cdots P_{i+1} G_i \tag{4.9}$$

P_i^j expresses the propagation of a carry through bits i through j, and G_i^j expresses the generation of a carry in any one of bits i to j and the subsequent propagation of that carry through the remaining bits in that block. Carries within blocks are still rippled, as before, but all other carries are now produced at the superblock level, using these composite signals.

For example, for a 32-bit adder of 4-bit blocks and 4-block superblocks, the relevant logic equations are

$$P_0^3 = P_3 P_2 P_1 P_0$$
$$P_4^7 = P_7 P_6 P_5 P_4$$
$$P_8^{11} = P_{11} P_{10} P_9 P_8$$
$$P_{12}^{15} = P_{15} P_{14} P_{13} P_{12}$$

$$P_{16}^{19} = P_{19} P_{18} P_{17} P_{16}$$
$$P_{20}^{23} = P_{23} P_{22} P_{21} P_{20}$$
$$P_{24}^{27} = P_{27} P_{26} P_{25} P_{24}$$
$$P_{28}^{31} = P_{31} P_{30} P_{29} P_{28}$$

$$G_0^3 = G_3 + P_3 G_2 + P_3 P_2 G_1 + P_3 P_2 P_1 G_0$$
$$G_4^7 = G_7 + P_7 G_6 + P_7 P_6 G_5 + P_7 P_6 P_5 G_4$$
$$G_8^{11} = G_{11} + P_{11} G_{10} + P_{11} P_{10} G_9 + P_{11} P_{10} P_9 G_8$$
$$G_{12}^{15} = G_{15} + P_{15} G_{14} + P_{15} P_{14} G_{13} + P_{15} P_{14} P_{13} G_{12}$$

$$G_{16}^{19} = G_{19} + P_{19}G_{18} + P_{19}P_{18}G_{17} + P_{19}P_{18}P_{17}G_{16}$$
$$G_{20}^{23} = G_{23} + P_{23}G_{22} + P_{23}P_{22}G_{21} + P_{23}P_{22}P_{21}G_{20}$$
$$G_{24}^{27} = G_{27} + P_{27}G_{26} + P_{27}P_{26}G_{25} + P_{27}P_{26}P_{25}G_{24}$$
$$G_{28}^{31} = G_{31} + P_{31}G_{30} + P_{31}P_{30}G_{29} + P_{31}P_{30}P_{29}G_{28}$$

$$C_3 = G_0^3 + P_0^3 C_{-1}$$
$$C_7 = G_4^7 + P_4^7 G_0^3 + P_4^7 P_0^3 C_{-1}$$
$$C_{11} = G_8^{11} + P_8^{11} G_4^7 + P_8^{11} P_4^7 G_0^3 + P_8^{11} P_4^7 P_0^3 C_{-1}$$
$$C_{15} = G_{12}^{15} + P_{12}^{15} G_8^{11} + P_{12}^{15} P_8^{11} P_4^7 G_0^3 + P_{12}^{15} P_8^{11} P_4^7 P_0^3 C_{-1}$$

$$C_{19} = G_{16}^{19} + P_{16}^{19} C_{15}$$
$$C_{23} = G_{20}^{23} + P_{20}^{23} G_{16}^{19} + P_{20}^{23} P_{16}^{19} C_{15}$$
$$C_{27} = G_{24}^{27} + P_{24}^{27} G_{20}^{23} + P_{24}^{27} P_{20}^{23} G_{16}^{19} + P_{24}^{27} P_{20}^{23} P_{16}^{19} C_{15}$$
$$C_{31} = G_{28}^{31} + P_{28}^{31} G_{24}^{27} + P_{28}^{31} P_{24}^{27} P_{20}^{23} G_{16}^{19} + P_{28}^{31} P_{24}^{27} P_{20}^{23} P_{16}^{19} C_{15}$$

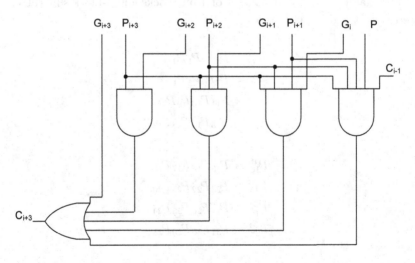

Figure 4.9: BCLA carry-lookahead

The high-level design of such an adder is shown in Figure 4.10. Similarly, a second level of lookahead may be used to improve the performance of the

RCLA when n is large. In this case, the function of the extra level of lookahead is to reduce the time required for the inter-block rippling. We leave it to the reader to work out the details of the relevant logic equations.

The formulations above (and others of a similar nature) of carry-lookahead adders are not entirely well-suited to current VLSI ASIC technology, although some may be for FPGA technology. For the former technology, there are better formulations, in the form of *parallel-prefix adders*, in which the signals P_i^j and G_i^j play a fundamental role. These adders are discussed below and are especially important as high-performance adders.

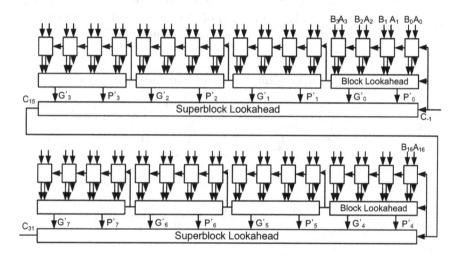

Figure 4.10: 32-bit Superblock BCLA

4.1.4 Conditional-sum adder

The design of the conditional-sum adder is based on the observation that, relative to the start of an addition, the carry bits (and, therefore, the sum bits) at the high end of the adder take longer to form than those at the low end. So fast addition can be achieved if at the high end of the adder two sets of sum bits are formed, one under the assumption of 0 carries and the other under the assumption of 1 carries, and the correct bits then quickly selected as soon as the correct carries are known.

Suppose an n-bit addition is divided into two $n/2$-bit blocks and two sets of sum bits formed for each block. When the carry out of the lower block is known, the correct sum bits in the upper block can be immediately determined. To form the sum and carry bits in the $n/2$-bit blocks, each is divided into two $n/4$-bit blocks, and the process outlined above is repeated. Proceeding in this manner, after $\log_2 n$ steps, we end up with n 1-bit blocks. At this point, assuming the carry into the adder is known, the selection process can begin at the least significant bits and proceed in reverse order. Thus the total number of sum bits determined at the various steps are one, two, four, and so forth. A conditional-sum adder therefore consists of $1 + \log_2 n$ levels (here numbered 0, 1, 2, ...) in which the block-sizes at each level are double those in the preceding level, and at the end of each level i, the number of final sum bits determined is 2^i. We shall use $_jS_i^q$ and $_jC_i^q$ denote the values at level j of bit i of the sum and bit i of the carry, respectively, under the assumption that the carry-in to bit i is q.

Table 4.1: 8-bit conditional-sum addition

i	7		6		5		4		3		2		1		0	
A	1		0		0		1		0		1		0		1	
B	0		0		1		0		1		1		0		1	
	C	S	C	S	C	S	C	S	C	S	C	S	C	S	C	S
C=0	0	1	0	0	0	1	0	1	0	1	1	0	0	0	**1**	**0**
C=1	1	0	0	1	1	0	1	0	1	0	1	1	0	1		
C=0	0	1		0	0	1		1	1	0		0	**0**	**1**		**0**
C=1	0	1		1	1	0		0	1	0		1				
C=0	0	1		0	0	1		1	**1**	**0**		**0**		**1**		**0**
C=1	0	1		1	1	0		0								
C=0	**0**	**1**		**1**		**0**		**0**		**0**		**0**		**1**		**0**

Table 4.1 shows an example of a conditional-sum addition. The bits selected at each level are shown in bold font. At level 0, the prospective sum and carry outputs are formed for each bit position. For the first bit-position, the carry-in is known (to be 0)[5] and, therefore, the correct sum and carry bits (S_0 and C_0) can be computed right away. At level 1, C_0 is known, and S_1 and C_1 can be computed; so at the end of this level, S_0 and S_1 are available. For all the other bit-positions at level 1, the correct (i.e. final) incoming carries are not known. So the 1-bit blocks from level 0 are combined into 2-bit blocks, with two sets of sum and carry outputs. The 2-bit blocks are formed as follows. Let i and $i+1$ be the bit-positions of the pair that end up in a block. Since the carry into block (i.e. into position

---

[5] We are here assuming unsigned addition.

i) is not known, S_i^0 and S_i^1 must be carried over from level 0. For position $i+1$, if ${}_0C_i^0 = {}_0 C_i^1$, then C_i is definitely known, and the correct sum bit is selected from level 0: if ${}_0C_i^0 = {}_0 C_i^1$, then ${}_1S_{i+1}^0 = {}_1 S_{i+1}^1 = {}_0 S_{i+1}^0 = {}_0 S_{i+1}^1$. On the other hand, if ${}_0C_i^0 \neq {}_0 C_i^1$, then, again, both possibilities from level 0 must be carried over: ${}_1S_{i+1}^q = {}_1 S_{i+1}^q$ and ${}_1C_{i+1}^q = {}_1 C_{i+1}^q$, where $q = 0, 1$. A similar procedure is carried out at level 2, with 4-bit blocks; at level 3, with 8-bit blocks; and so on, until all the sum bits, S_i, have been produced.

Figure 4.11 shows the high-level organization for an 8-bit conditional-sum adder. The relevant logic equations are (for $C_{-1} = 0$)

$$C_0 = A_0 B_0$$
$$S_0 = A_0 \oplus B_0$$
$${}_0C_i^0 = A_i B_i \qquad i = 1, 2, \ldots, n-1$$
$${}_0C_i^1 = A_i + B_i$$
$${}_0S_i^0 = A_i \oplus B_i$$
$${}_0S_i^1 = \overline{A_i \oplus B_i}$$

for level 0. For each subsequent level, k, the sum and carry bits that are completely determined are

$$S_i = {}_{k-1}S_i^0 \; {}_{k-1}\overline{C}_{i-1} + {}_{k-1}S_i^1 \; {}_{k-1}C_{i-1} \quad i = 2^{k-1}, 2^{k-1}+1, \ldots, 2^k - 1$$
$$C_{2^k-1} = {}_{k-1}C_{2^k-1}^0 \; {}_{k-1}\overline{C}_{i-1} + {}_{k-1}C_{2^k-1}^1 \; {}_{k-1}C_{i-1}$$

and for the sum bits that are unconditionally carried over (at the positions where level $k-1$ blocks coalesce)

$$_kS_q^q = {}_{k-1}S_i^q \qquad i = 2^k, 2^{k+1}+1, 2^{k+3}, \ldots \quad \text{and} \quad q = 0, 1$$

For the remaining bit positions, i,

$$_kS_i^q = {}_{k-1}S_i^q \; {}_{k-1}\overline{C}_{i-1} + {}_{k-1}S_i^q \; {}_{k-1}C_{i-1}^q \qquad q = 0, 1$$
$$_kC_i^q = {}_{k-1}C_i^q \; {}_{k-1}\overline{C}_{i-1} + {}_{k-1}C_i^q \; {}_{k-1}C_{i-1}^q$$

A logic circuit for level-1 modules is shown in Figure 4.12. Close examination of the equations above shows that they are just equations for 2-to-1 multiplexers; and, indeed, they are sometimes realized as exactly that. The logic equations also reveal the main drawback of the conditional-sum adder: the fan-out required of the C_is doubles with each level, and the distances over with they must be transmitted also increase. Consequently, for large n, the conditional-sum technique is almost always used in combination with some other technique in which it is limited to small blocks.

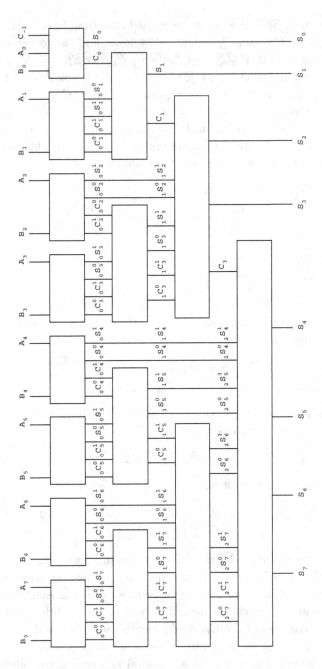

Figure 4.11: 8-bit conditional-sum adder

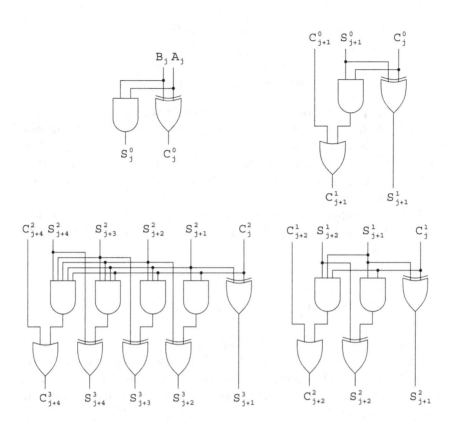

Figure 4.12: Level-1 logic for conditional-sum adder

4.1.5 Parallel-prefix adders

Parallel-prefix adders allow more efficient implementations of the carry-lookahead technique and are, essentially, variants of the two-level carry-lookahead adders of Section 4.1.3. Indeed, in current technology, parallel-prefix adders are among the best adders, with respect to *area× time* (*cost:performance ratio*), and are particularly good for the high-speed addition of large numbers. For residue arithmetic, they also have the advantage, as we shall see below, of being easily modifiable for addition with respect to the special moduli $2^n - 1$ and $2^n + 1$. In what follows, we shall define the *prefix problem*, show that the computation of carries can be reduced to this, and then discuss the varieties of parallel-prefix adders.

Let $S \stackrel{\triangle}{=} \{a_{n-1}, a_{n-2}, \ldots, a_0\}$ be a set that is closed under an associative operator, $\bullet$. For $a_i \in S$ the problem of computing the *prefixes*

$$x_i = x_{i-1} \bullet a_i \qquad 1 \le i \le n-1$$
$$x_0 = a_0$$

is called the prefix problem. Many important problems in computing can be shown to be equivalent to that of computing prefixes [11]. In the case of adders, the computation of carries is equivalent to the prefix problem, as we next show. The most important aspect of the prefixes is that they, and their subterms, can be computed in parallel. This follows from the property of associativity.

Before we define the operator $\bullet$ for the computation of carries, first observe that the block carry-generate signal G_i^j (Equation 4.9) may be expressed as

$$G_i^j = \begin{cases} G_i & \text{if } i = j \\ G_{k+1}^j + P_{k+1}^j G_i^k & \text{if } i \le k < j \end{cases}$$

This expression indicates the different ways (and with varying degrees of parallelism) in which the bit-level P_i and G_i signals may be combined to get the corresponding block signals. For example, the basic expression for G_0^3 is $G_3 + P_3 G_2 + P_3 P_2 G_1 + P_3 P_2 P_1 G_0$, and this may be evaluated in different ways, such as

$$G_3 + P_3(G_2 + P_2 G_1 + P_2 P_1 G_0) = G_3^3 + P_3^3 G_0^2$$

$$(G_3 + P_3 G_2) + P_3 P_2 (G_1 + P_1 G_0) = G_2^3 + P_2^3 G_0^1$$

$$(G_3 + P_3 G_2 + P_3 P_2 G_1) + P_3 P_2 P_1 G_0 = G_1^3 + P_1^3 G_0^0$$

Similarly, the block carry-propagate signal (Equation 4.9) may be expressed as

$$P_i^j = \begin{cases} P_i & \text{if } i = j \\ P_{k+1}^j P_i^k & \text{if } i \le k < j \end{cases}$$

As in the preceding case, this allows for some variation in the order of evaluation. For example, some of the ways in which P_0^3 may be evaluated

are
$$(P_3P_2P_1)P_0 = P_1^3 P_0^0$$

$$(P_3P_2)(P_1P_0) = P_2^3 P_0^1$$

$$P_3(P_2P_1P_0) = P_3^3 P_0^2$$

We now define the operator • as follows.
$$\left[G_{k+1}^j, P_{k+1}^j\right] \bullet \left[G_i^k, P_i^k\right] \stackrel{\triangle}{=} \left[G_{k+1}^j + P_{k+1}^j G_i^k, P_{k+1}^j P_i^k\right]$$
$$= \left[G_i^j, P_i^j\right]$$

It is shown easily that • is associative. This means that the subterms that form given block-generate and block-propagate signals can be computed in parallel, with the degree of parallelism determined by the selection of the subterms involved. The same remark applies to the computation of carries: if the carry-in, C_{-1}, to the adder is 0, then (by Equations 4.8 and 4.9)

$$C_i = G_0^i \qquad\qquad 0 \le i \le n$$
$$= G_j^i + P_j^i G_0^{j-1} \qquad 1 \le j \le i-1$$

Note that the operator • produces both block carry-propagate and carry-generate signals; but at the last level, where carries are produced, the latter are not necessary. We shall use ○ to denote the prefix operators at that level.

We have seen above (Section 4.1.3) that a carry may also be propagated by using *kill* signals instead of *transfer* or *propagate* signals. The formulations above can therefore be replaced with ones based on K_i or T_i. For example, for K_i, we have

$$\overline{K}_i^j = \begin{cases} \overline{K}_i & \text{if } i = j \\ \overline{K}_{l+1}^j \overline{K}_i^j & \text{if } i \le l < j \end{cases}$$

$$G_i^j = \begin{cases} G_i & \text{if } i = j \\ G_{l+1}^j + \overline{K}_{l+1}^j G_i^k & \text{if } i \le l < j \end{cases}$$

$$\left[G_{l+1}^j, \overline{K}_{l+1}^j\right] \bullet \left[G_i^l, \overline{K}_i^l\right]$$
$$\stackrel{\triangle}{=} \left[G_{l+1}^j + \overline{K}_{l+1}^j G_i^l, \overline{K}_{l+1}^j \overline{K}_i^l\right]$$
$$= \left[G_i^j, \overline{K}_i^j\right]$$

This is common in many realizations of parallel-prefix adders. We shall, however, continue to use just the P signals.

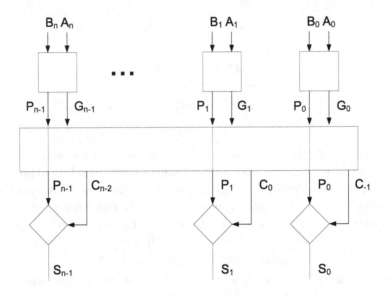

Figure 4.13: Generic parallel-prefix adder

In addition to associativity, • has another useful property: idempotency. That is

$$\left[G_l^j, P_l^j\right] \bullet \left[G_i^k, P_i^k\right] = \left[G_i^j, P_i^j\right] \qquad i \leq l \leq k < j$$

For example, it is easy to show that

$$\left[G_3^7, P_3^7\right] \bullet \left[G_2^4, P_2^4\right] = \left[G_2^7, P_2^7\right]$$

What idempotency means is that the blocks need not be contiguous: there may be some overlap. This is useful because it enhances the ways in which parallelism can be exploited.

The generic structure of a parallel-prefix adder is shown in Figure 4.13. A □ represents the logic that produces bit-level carry propagate and generate signals; and a ◊ represents sum-formation logic. Parallel-prefix adders differ primarily in the details of the carry-prefix tree, in which the block-generate and block-propagate signals, as well as the carries, are formed. By varying the selection of sub-terms in the corresponding full expressions of the block signals, and, therefore, varying the fan-in and fan-out requirements, as well as the length of the interconnections, different adders are obtained that span the cost-performance design-space. One aspect that all such adder-designs have in common is that the operational time of all the highly parallel adders is $O(\log n)$, which is the number of levels in the prefix tree, although the constants involved can vary substantially.

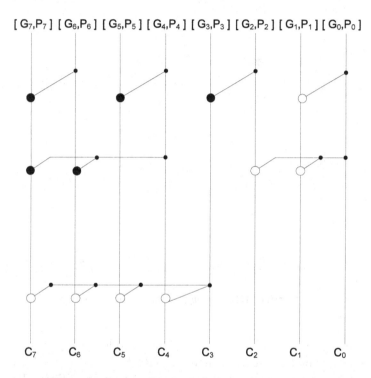

Figure 4.14: 8-bit Ladner-Fischer prefix tree

The best-known early parallel-prefix adder is the Ladner-Fischer adder, whose prefix tree has the form shown in the example of Figure 4.14 [16]. The black and empty circles in the diagram correspond to the operators, • and ∘, defined above. One drawback of this design is that the lateral fan-out required of the prefix cells doubles at every level: one, two, four, eight, etc. Thus in a realization, buffers might be required to provide additional drive and these, as they will be on the adder's critical path, can adversely affect performance. The Ladner-Fischer adder makes use of the associativity of •

but not its idempotency.

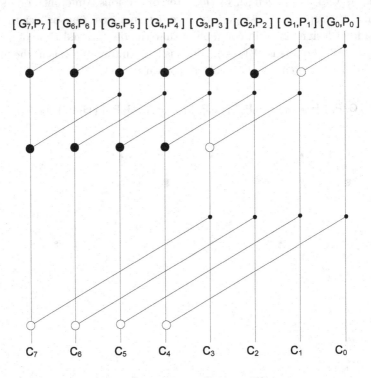

Figure 4.15: Kogge-Stone prefix tree

The Kogge-Stone adder is a parallel-prefix adder in which the lateral fan-out of the prefix cells is limited to unity [17]. Figure 4.15 shows a Kogge-Stone prefix-tree, in which two things are evident. First, the number of interconnections is much higher than in the Ladner-Fischer adder, and they are also longer. This is significant because in current technology interconnections have a significant effect on area and operational time. Second, more prefix cells are used. On the whole, however, Kogge-Stone adders tend to be slightly faster than Ladner-Fischer adders, but at the expense of requiring more area.

The Brent-Kung adder is another parallel-prefix adder in which all lateral fan-out is limited to unity [2]. An example prefix-tree is shown in Figure 4.16. The adder is, however, not the best for very fast addition: in current VLSI technology, a substantial part of the operational delay will be due to the lengths of the interconnections, and the critical path in this adder is rather long.

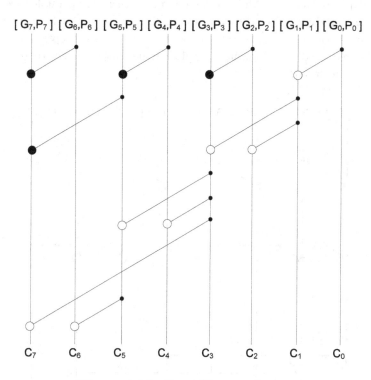

Figure 4.16: Brent-Kung prefix tree

The most recent major advance in the design of parallel-prefix adders has been the work of Knowles [18]. Essentially, this work has unified a wide range of parallel-prefix adders and in the process revealed several previously unknown ones. The basic idea is that different adders can be obtained by varying fan-outs and the number/lengths of interconnections in the prefix tree. If we represent adders by their lateral fan-outs, then the Ladner-Fischer adder is a [1, 2, 4, 8, ...] adder, and the Kogge-Stone adder is a [1, 1, 1, 1, ...] adder. What the new work has shown is that there are several other adders between these two endpoints in the design-space. For example, for 8-bit addition, the Ladner-Fischer adder has a [1, 2, 4] prefix tree, that of the Kogge-Stone adder is [1, 1, 1], but other possibilities include [1, 1, 2], [1, 1, 4], and [1, 2, 2]. The last of these is shown in Figure 4.17.

So far, we have assumed that prefix cells have a fan-in of two. It is natural to consider varying this fan-in, as a higher fan-in means a decrease in the depth of the tree. An increase in fan-in is certainly possible, because of the associativity of •. But in practice, the fan-in cannot be arbitrarily increased, because there is a trade-off between the reduced depth of the tree and the increased delay through a cell. Further discussion of these issues will be found in [1].

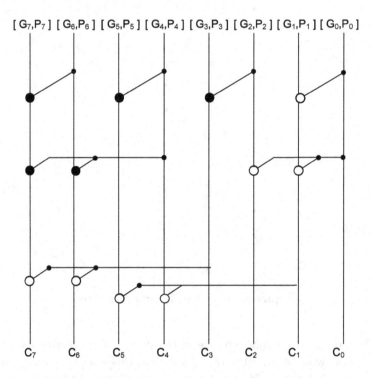

Figure 4.17: [1, 2, 2] prefix adder tree

4.1.6 Carry-select adder

In the conditional-sum adder, we start with 1-bit blocks (adders), produce both possible sum (and carry) bits at every bit-position, and then go though a selection tree in which the correct sum bits are determined. At every level of the tree, the block-sizes and the number of final sum-bits produced are double those at the preceding level. In a sense, though, the sizes of both the initial blocks and those of blocks at succeeding levels are arbitrary: there is no reason why we should not choose different sizes at any level, even allowing for the fact that uniformity implies regularity in the realization. This is exactly what is done in the carry-select adder, of which the conditional-sum adder may be viewed as just one instance. Indeed, the conditional-sum adder is a "pure" carry-select adder, in contrast with the "hybrids" that we next describe. The underlying motivation is the same—to take advantage of the fact that high-order carry and sum bits take longer to form than low-order ones.

The generic form of a carry-select adder with m-bit blocks is shown in Figure 4.18. Each block at the initial level consists of two m-bit adders: one that produces sum bits under the assumption of a 0-carry into the block and another that produces sum bits under the assumption of a 1-carry. When the carry-out of one block is known, it is used to select one set of sum bits in the next block. We shall use S_i^q and C_i^q to denote the values at bit-slice i of the sum and carry, respectively, under the assumption that the carry into the block containing bit i is q. Note that if C_{-1} is known —and we usually take it to be 0, for unsigned addition —then the least significant m-bit adder takes C_{-1} as an additional input and produces the least significant m bits of the final sum, as well as the carry out of the block.

The general equations for an m-bit block, j ($j = 0, 1, 2, \ldots$), with block carry-in C_{j-1} and block carry-out C_{j+m} are

$$S_i = S_i^0 \overline{C}_{j-1} + S_i^1 C_{j-1} \qquad i = jm, jm+1, \ldots j(m+1) - 1$$
$$C_{j+m} = C_{j+m}^0 \overline{C}_{j-1} + C_{j+m}^1 C_{j-1}$$

which, again, are just the equations for 2-to-1 multiplexers. The m-bit adders may be of any of the designs already discussed above: ripple, carry-skip, carry-lookahead, and so forth. The carry-select technique therefore leads to a whole family of "hybrid" adders, covering a range of performance:cost ratios.

Although, we have implied that each block of a carry-select adder consists of two separate adders, it should be noted that the actual logic required need not be twice that of a single adder. Depending on the design of the underlying adders, some sharing of logic is possible. For example, if carry-lookahead is used, then the carry-propagate and carry-generate logic will be the same for both adders. Tyagi has used this observation to design a carry-select/parallel-prefix adder, in which the sharing reduces the cost by about three gates per bit-slice [19].

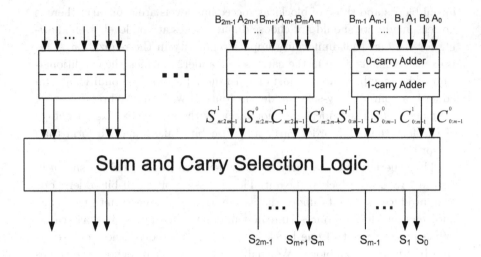

Figure 4.18: Generic carry-select adder

The formulations above for the parallel-prefix adders (Section 4.1.5) assume the carry into the adder is 0. For a carry-select/parallel-prefix adder, in principle, another prefix tree is needed for a carry-in of 1. If we define $\widetilde{P}_i^j$ and $\widetilde{G}_i^j$ to be the block propagate and generate signals (Equations 4.8 and 4.9) under the assumption of a 1-carry, then it can be shown that all of the formulations of Section 4.1.5 go through with these new signals in place of P_i^j and G_i^j. It can also be shown that

$$[\widetilde{G}_i^j, \widetilde{P}_i^j] = [G_i^j + P_i^j, P_i^j]$$

which allows us to easily obtain C_i^1 from C_i^0:

$$C_i^1 = C_i^0 + P_0^i$$

Another point to note is that although Figure 4.18 shows blocks of the same size, block sizes may differ, with some advantage. Consider two adjacent blocks j and $j+1$. The prospective sum bits in block $j+1$ will become available some time before the carry into the block. Therefore, block $j+1$ may be made slightly larger than block j without any loss in performance. Recursively carrying out this line of reasoning, from the low end of the adder to the high end, we may conclude that block-sizes should increase, disproportionately, from right to left, with the aim of making equal the time required to generate the sum bits at a given bit-slice and the time required to get the correct carry into that bit-slice. The adders in [19], for example, use the block-sizes $1, 1, 2, 4, 8, \ldots, n/2$. (Of course, the best sizes will depend on the design of the underlying adders and the realization-technology used.) Tyagi's design in fact yields, by varying the block-sizes, a range of adders that span the entire space between the $O(\sqrt{n})$ operational time of the carry-skip adders and the $O(\log n)$ time of the parallel-prefix adders. A different deign for a hybrid carry-select adder is described in [9]. This one combines carry-selection, carry-lookahead, and Manchester carry-chains.

4.2 Residue addition: arbitrary modulus

In this section, we shall discuss addition relative to moduli other than the special moduli 2^n, $2^n - 1$, and $2^n + 1$; addition with the latter three moduli is discussed in Sections 4.3 and 4.4. When the modulus is not constrained, it is rather difficult to design an efficient residue adder; any residue adder will be substantially slower or more costly than a conventional adder of the same word-length.

We shall begin by discussing high-level designs for modulo-m adders, and we shall initially assume that m may be varied. Subsequently, we shall consider the optimization of the residue adders, according to which of the underlying adders of Section 4.1 are used. The most worthwhile of such optimizations rely on the fact that the modulus, even if it is not one of the special ones, is usually fixed, i.e. known at design-time.

The result of adding, modulo-m, two numbers, A and B, (digits of a residue representation), where $0 \leq A, B < m$, is defined as

$$|A + B|_m = \begin{cases} A + B & \text{if } A + B < m \\ A + B - m & \text{otherwise} \end{cases} \quad (4.10)$$

The design of a modulo-m adder may be based, directly or with some variation, on Equation 4.10. The most straightforward implementation is one that operates as follows. A and B are added to yield an intermediate sum, S'. S' is then compared with m. If S' is less than m, then S' is the result; otherwise, m is subtracted from S', to yield a new result, S'', which is then the correct result. (The subtraction may be implemented by adding the two's complement of m, which complement we shall denote by $\overline{m}$.)

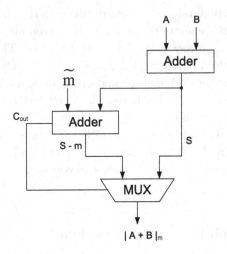

Figure 4.19: Basic modulo-m adder

Implementing this procedure in a simplistic way requires three (carry-propagate) adders—one for the addition, one for the subtraction, and one for the comparison. But closer examination shows that the comparison and the second addition (i.e the subtraction) can be combined. First, A and B are added. Next, the subtraction (i.e. the addition of $\overline{m}$, The twos complement of m) is immediately carried out. A carry-out[6] from the second addition then indicates whether or not S' is less than m: if there is a carry-out, then the correct result is S''; otherwise, it is S'.[7] The corresponding hardware organization is shown in Figure 4.19, in which the adders may be of any of the designs of Section 4.1.

A less costly, but slower modulo-m adder, can be obtained by modifying the adder of Figure 4.19 so that it uses one adder but instead operates in two cycles. The resulting design is shown in Figure 4.20. During the first cycle, A and B are added and the result, S', latched. In the second cycle, m is subtracted, to form S''. Then, as before, one of these two intermediate results is selected to be the final result. An example of this type of adder

---

[6] The sign bit could also be used instead.

[7] The justification for using the carry-out is as follows. When interpreted as an unsigned number, the n-bit representation of $\overline{m}$ corresponds to the numerical value $2^n - m$ (see Chapter 1). So if $A + B \geq m$, then $A + B + \overline{m} = (A + B - m) + 2^n$. $A + B - m \geq 0$; so the 2^n is a carry out of the most significant bit-slice. But if, $A + B < m$, then $A + B + \overline{m} < 2^n$, and there is no carry-out.

is described in [22]. If, on the other hand, faster addition is required, then a straightforward way to achieve this is to concurrently compute both S' and S'' and then select one. The computation of S'' requires that three operands be reduced to one, which is easily done by using a combination of a carry-save adder (CSA) and a carry-propagate adder (CPA). The final modulo-m adder design is then as shown in Figure 4.21. Essentially, what we have here is a carry-select adder that computes $A+B/A+B-m$ instead of the conventional $A+B/A+B+1$. It should, however, be noted that the speed advantages of this new adder are not as substantial as might initially appear. This is because the two adders in Figure 4.19 can run in parallel for most of the time: the critical path consists of the operational time of the second adder plus the time through one bit-slice of the first adder. Of course, the actual benefits depend on exactly how the adders are implemented: if, for example, they are ripple adders, then the performance improvement would be, relative to the extra cost, of dubious worth.

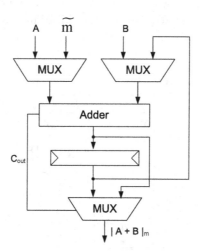

Figure 4.20: Reduced-cost modulo-m adder

It should be noted that although in Figure 4.21 we have shown two nominal CPAs, in practice, depending on the underlying design of a CPA, some sharing of logic is possible. Therefore, the total cost need not be twice that of a single adder.

If the modulus m is of sufficiently small (i.e. of low precision), and, therefore, the two operands also are, then the modulo-m addition can be reasonably realized through a look-up table. Such a table would nominally use the concatenation of the two operands, A and B, as an address, and for the output it would return the result $|A+B|_m$. Evidently, advantage may be taken of symmetry to reduce the basic table-size by half. A less direct variant of the table-lookup method involves storing just the first row of the nominal addition table and making use of this together with some combinational logic (a shifter and decoder). The basis of this design is the observation that any other row, k, of the addition table can be obtained by a k-place rotation (cyclic shift) of the elements in the first row; see Table 4.2 for example.[8] Thus the addition of A and B, modulo m, may be carried out by performing a rotation by A places and then decoding B to select an element of the result of the rotation. This leads to the design of Figure 4.22. The trade-off in such a design, relative to the above approaches, may be a reduction in operational delay but at the price of an increase in logic, and for large moduli the logic required for the shifter may render this approach uncompetitive.

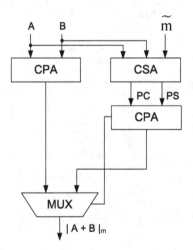

Figure 4.21: Enhanced-performance modulo-m adder

---

[8]The theoretical underpinnings for this will be found in Chapters 1 and 2.

Table 4.2: Modulo-5 addition table

A\B	0	1	2	3	4
0	0	1	2	3	4
1	1	2	3	4	0
2	2	3	4	0	1
3	3	4	0	1	2
4	4	0	1	2	3

We now consider the optimization of the three high-level designs above (Figures 4.19, 4.20, and 4.21). There is not much room for improvements in the design of Figure 4.20. The only worthwhile one might be the use of logic that combines both computational and storage functions but without introducing additional operational delay. On the other hand, the other two designs offer more opportunities for improvement, depending on what the underlying adders are. We shall briefly examine the various possibilities, according to each of the designs of Section 4.1. We shall assume that each adder is of word-length n bits, where $n = \lceil \log_2 m \rceil$ and m is the modulus.

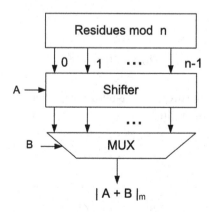

Figure 4.22: Table-derived modulo-m adder

Regardless of the design of the underlying adder, the top adder in Figure 4.19 (which is functionally the same as the left adder in Figure 4.21) and the CSA in Figure 4.21 will produce signals that are equivalent to the carry-propagate (P_i) and carry-generate and (G_i) signals of Section 4.1.3. So in the design of Figure 4.21, these signals may be shared. This will reduce the cost by n full adders but without an increase in the operational time. One can think of other optimizations of this nature, but it is doubtful that any will be worthwhile if the modulus is not fixed (i.e. known at design-time).

If m is known at design-time, and it usually is, then there is an obvious and fruitful way to optimize both the designs of Figures 4.19 and 4.21: take each bit-slice, for bit $\overline{m}_i$, of $\overline{m}$ and optimize it according to whether $\overline{m}_i$ is 0 or 1. Generally, the bit-slices for $\overline{m}_i = 0$ will be simpler than those for $\overline{m}_i = 1$. That might make a difference when cost and performance are measured crudely, say by gate-count and gate delays, but for more accurate measures (e.g. area and raw time), the difference might not be appreciable. Still, trying to cope with the irregularity that arises from having different bit-slices gives us additional insight (beyond those of Chapter 1) in how to choose moduli: try to ensure that the representation of $\overline{m}$ has as many 0s as possible; that is, that of m has as many 1s as possible.

Let us now consider the case when the underlying adders are ripple adders. In this case a bit-slice will consists of two full adders (and a multi-plexer), for which the equations are (see Equations 4.1 and 4.1)

$$S'_i = (A_i \oplus B_i) \oplus C_{i-1}$$
$$C_i = A_i B_i + (A_i \oplus B_i) C_{i-1}$$

$$S'_i = (\overline{m}_i \oplus S_i) \oplus C_{i-1}$$
$$C'_i = \overline{m}_i S_i + (\overline{m}_i \oplus S_i) C_{i-1}$$

with S_i then selected through a 1-bit multiplexer. If $\overline{m}_i = 0$, then the second set of equations reduces to

$$S'_i = S_i \oplus C_{i-1}$$
$$C'_i = S_i C_{i-1}$$

And if $\overline{m}_i = 1$

$$S'_i = \overline{\oplus S_i} \oplus C_{i-1}$$
$$C'_i = S'_i + \overline{S_i} C_{i-1}$$

Addition 117

For the residue adder in Figure 4.21, the changes are more interesting, even for a ripple adder. For the case where the CPA is a ripple adder, the original bit-slice is shown in Figure 4.23. If $\overline{m}_i = 0$, then

$$PC'_i = G_i$$
$$PS'_i = P_i$$

where PC_i and PS_i are the partial (i.e. unassimilated) carry and sum bits from adding A_i and B_i. Relative to Figure 4.23, one full adder can be omitted.

And if $\overline{m}_i = 1$, then

$$PC'_i = P_i + G_i = T_i$$
$$PS'_i = \overline{P}_i$$

By sharing the logic for G_i/P_i and introducing a new gate for T_i, we end up with the new bit-slice shown in Figure 4.24. We shall assume these designs in discussions with respect to the other adders in Section 4.1. The relevant changes for the design of Figure 4.19 are largely similar and are left as an exercise for the reader.

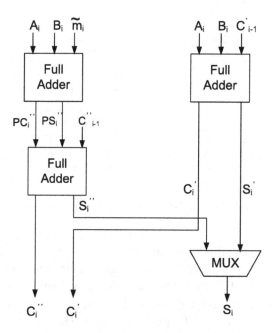

Figure 4.23: Original bit-slice of modulo-m ripple adder

There is an additional "trick" that can be useful, depending on the adder. In Figure 4.24 (and, in general, Figure 4.21), both left and right adders have similar components. So the same hardware can be used for both computations, if an appropriate choice is made between the inputs. Thus, instead of choosing between sum bits, the multiplexer may be moved up, to select the propagate and generate bits instead. C''_{n-1}, the carry out of the second adder, now needs to be computed earlier; and the best way to do this is to use carry-lookahead (for just that one carry bit), irrespective of the design of the underlying adder. An example of this type of optimization is given below, with a carry-lookahead adder as the underlying adder.

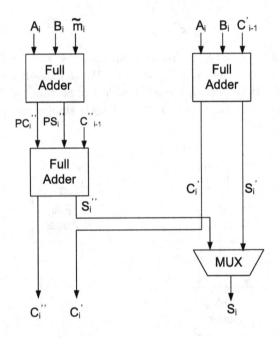

Figure 4.24: Optimized 1-bit-slice slices of modulo-m ripple adder

The design of a residue carry-skip adder based on Figure 4.23 is straightforward and involves the use two carry-skip chains. One carry-skip chain is similar to that of Figure 4.5. The other consists of a combination of optimized 0-slices and 1-slices, and in Equation 4.2, P_i is replaced with $P_i \oplus G_{i-1}$, if $\overline{m}_i = 0$, and with $P_i \oplus T_{i-1}$, if $\overline{m}_i = 1$, where $P_i = A_i \oplus B_i$, $G_i = A_i B_i$, and $T_i = A_i + B_i$. Similarly, for a carry-lookahead adder we would have two carry networks, which collectively would replace the bottom half-adder and OR gate in each side if Figure 4.26. The first network is exactly as described by Equations 4.4 to 4.17. For the second network, if $\overline{m}_i = 0$, then P_i and G_i in those equations are replaced with

$$\widehat{P}_i \triangleq P_i \oplus G_{i-1}$$
$$\widehat{G}_i \triangleq P_i G_{i-1}$$

and $\overline{m}_i = 1$, then the replacements are

$$\widehat{P}_i = \overline{P}_i \oplus T_{i-1}$$
$$\widehat{G}_i = \overline{P}_i T_{i-1}$$

Let us now consider the design of a modulo-m adder, based on, say, Figure 4.21. Two types of optimizations are possible. The first consists of optimizing each bit-slice of the adder, according to the value of m_i, as above. The second type largely involves a reduction in costs. A straightforward design based on Figure 4.21 would involve the use of two lookahead networks. Now, most of the logic in a carry-lookahead adder is taken up by the carry-network. It would therefore be advantageous if only one carry-network were used, and this can be done by moving up the multiplexer, to select the propagate and generate bits instead of the sum bits. C_{n-1}, the carry out of the adder, must now be produced outside the rest of the carry-network. We leave it to the reader to work out the details for a specific modulus. (An adder of this type is described in [23].) The new adder will cost much less than a straightforward carry-lookahead version of Figure 4.21, but it will also be slightly slower.

Modulo-m parallel-prefix adders are easy to obtain from the last design: simply replace a carry network with a prefix-tree and optimize accordingly.

4.3 Addition modulo $2^n - 1$

The modulus $2^n - 1$ is an especially important one because modulo-$2^n - 1$ arithmetic units can be designed to be almost as efficient as corresponding ones for conventional arithmetic. We have seen, in Section 4.2, that to simplify the straightforward design of a residue adder, it helps to have a modulus whose complement-representation has as many 0s as possible; that is, the representation of the modulus has as many 1s as possible. For signed numbers, the representation (of a positive number) that has the greatest number of 1s is 011...1 (which represents $2^n - 1$). The two's complement of this is 100...01; and the one's complement is 100...0, which has more 0 than the former. It is therefore not surprising that excluding 2^n, for which residue arithmetic is just conventional arithmetic, $2^n - 1$ is the best modulus in terms of implementation and that the most efficient modulo-$(2^n - 1)$ adder is just a slightly modified one's complement adder.

In designing a modulo-$2^n - 1$ adder, it is useful to distinguish among three cases, depending on the intermediate result of the addition of the two operands, A and B, where $0 \leq A, B < 2^n - 1$:

- $0 \leq A + B < 2^n - 1$
- $A + B = 2^n - 1$
- $2^n - 1 < A + B < 2^{n+1} - 2$

In the first case, the intermediate result is the correct modulo-$(2^n - 1)$ result. In the second and third cases, we should subtract $2^n - 1$ in order to get the correct result, which subtraction is equivalent to subtracting 2^n and adding 1. The third case is the more straightforward of the latter two, both in determining whether or not the result is within the given range and in carrying out the required correction. An intermediate result greater than 2^n implies a carry out of the most significant bit-position of the adder; so the "excess $2^n - 1$" case is known to have occurred when there is a carry-out from the addition. Ignoring this carry-out is equivalent to subtracting 2^n; and adding 1 then produces the desired result. Correction in the second case too is easily dealt with by simply adding a 1. But detecting that the intermediate result is equal to $2^n - 1$ is not as straightforward as detecting when the modulus has been exceeded: there is no carry-out to indicate when the addition of a corrective 1 should be carried out. Nevertheless, it is not particularly difficult to realize, as we show below. Examples of all three cases are given in Table 4.3.

The addition process just described is evidently similar to one's complement addition: If $2^n - 1$ is accepted as another representation for zero—recall that one's complement has two representations for that number—then the process is exactly one's complement addition, and $2^n - 1$ is the other representation for zero. On the other hand, if we proceed exactly as above, and correct this "boundary" result, then addition modulo $2^n - 1$ is not equivalent to one's complement addition. Permitting two representations for zero is sometimes useful, e.g. in fault-tolerance applications of residue number systems, and it leads to a slightly better design for the residue adder. Nevertheless, for normal residue arithmetic, having both representations is not desirable, and, unless otherwise specified, we shall assume that there is to be only one representation for zero, i.e. $00\cdots0$.

Table 4.3: Examples of additions modulo $2^n - 1$

m = 15, n = 4

A	0100	(4)	
B	<u>0101</u>	(5)	
	1001	(9)	= 4 + 5 mod 15

A	0111	(7)	
B	<u>1000</u>	(8)	
	1111	(15)	
	<u>+ 1</u>	(correction)	
	0000	(0)	= 7 + 8 mod 15

A	1001	(9)	
B	<u>1010</u>	(10)	
	1011		
	<u>+ 1</u>	(end -around -carry)	
	0100	(4)	= 9 + 10 mod 15

Modulo-$(2^n - 1)$ addition may be implemented in a variety of ways, depending on the design of the underlying (conventional) adder and whether or not the same adder is used twice (for the primary addition for the corrective increment). Let us assume, for the moment, that two representations for zero are acceptable. A straightforward implementation of the residue adder would consist of a conventional adder and an incrementer, with the sum outputs of the adder providing the inputs to the incrementer and the carry-out of the adder providing the carry-in (or least-significant-bit input) for the incrementer. It is, however, possible to design a residue adder that has a smaller operational delay and requires less logic than such an arrangement. If one were to use a conventional adder for residue addition in two (nominal) cycles, as described in Section 4.2, then it may be observed that, depending on the adder design, the actions required during the first cycle and those required during the second cycle are mutually exclusive.

The consequence of this observation is that with just the inclusion of just a little more logic to the underlying adder, it is possible to obtain a residue adder that operates in a single cycle. We shall show this, in a most direct way, for the case of the carry-lookahead adder.

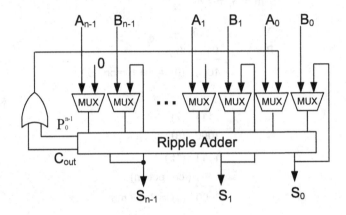

Figure 4.25: Modulo-$2^n - 1$ ripple adder

4.3.1 Ripple adder

It might appear straightforward to convert a conventional ripple adder into a modulo-$(2^n - 1)$ adder if two representations are allowed for zero: simply connect the carry-out line (from the most significant bit-slice) to the carry-in line (into the least significant bit-slice). In practice, however, the resulting feedback path can create a race condition between the two stable states of the adder, and this may require the allowance of several cycles for the outputs of the adder to stabilize. One way to get around this would be to use latches, so that the outputs of the adder are effectively latched before being fed back as inputs. If, on the other hand, only one representation for zero is permitted, then the signal that is used to indicate an intermediate result of $2^n - 1$ may also be used to eliminate the race condition [14, 15]. This signal is determined as follows. $C_\text{out} = 1$ indicates that the modulus has been exceeded. And a result of $11\ldots1$ occurs only when for every bit-slice, either $A_i = 0$ and $B_i = 1$ or $A_i = 1$ and $B_i = 0$; that is $P_i = 1$ for every i, and so $P_0^{n-1} = 1$. (See Section 4.1.3 for the definition of these signals.) Therefore, the required signal is $C_\text{out} + P_0^{n-1}$. The corresponding adder design is shown in Figure 4.25.

The design of Figure 4.25 can be easily modified into a residue carry-skip adder. In such a case the signal P_0^{n-1} is readily obtained from the block signals, P_j^{jm-1}, that are already required for the carry-skip chain.

4.3.2 Carry-lookahead adder

As an example of the approach described at the end of the introduction to this section, let us consider the case where the underlying adder is a one-level carry-lookahead adder (CLA), such as that shown in Figure 4.6. The logic equations for such an adder are

$$G_i = A_i B_i$$
$$P_i = \overline{A_i} B_i + A_i \overline{B_i} = A_i \oplus B_i$$
$$C_i = G_i + P_i C_{i-1}$$
$$S_i = P_i \oplus C_{i-1}$$

and unwinding the carry equation yields

$$C_0 = G_0 + P_0 C_{-1}$$
$$C_1 = G_1 + P_1 G_0 + P_1 P_0 C_{-1}$$
$$C_2 = G_2 + P_2 G_1 + P_2 P_1 G_0 + P_2 P_1 P_0 C_{-1}$$
$$\vdots$$
$$C_i = G_i + P_i G_{i-1} + P_i P_{i-1} G_{i-2} + \cdots + P_i P_{i-1} P_{i-2} \cdots P_0 C_{-1}$$

where C_{-1} is the carry-in to adder, i.e. into the least significant bit-position.

Suppose such an adder is used to perform residue addition in two cycles. The first cycle would consist of the addition of A and B, with $C_{-1} = 0$. And in the second cycle $C_{-1} = C_{\text{out}}$, where C_{out} is the carry-out from the first addition. Consider now the least significant bit-slice. Let C_i^q denote the carry from bit-slice i during the cycle q, let $S'_{n-1} S'_{n-2} \ldots S'_0$ denote the (intermediate) sum produced in that cycle, and let G_i and P_i be the carry generate and propagate signals during the addition (in the first cycle) of A and B. In the first cycle, since $C_{-1}^1 = 0$, and there is a carry from bit-slice 0 only if one is *generated* there; that is, $C_0^1 = G_0$. In the second cycle, the operands into that bit slice are S'_0, an implicit 0, and C_{-1} (formerly C_{n-1}^1). In that cycle a carry cannot be generated from bit-slice 0, but C_{-1} may be *propagated*; and the propagation signal in that case is $S'_0 \oplus 0$, which is just S'_0. Now, $S'_0 = P_0$ (since $C_{-1}^1 = 0$; so $C_{-1}^2 = P_0 C_{n-1}^1$. Note that C_0^1 and C_0^2 are independent, and the expression for a carry in either cycle is $C_0^1 + C_0^2 = G_0 + P_0 C_{-1}^2$, which is exactly the normal expression for a carry from that position (Equation 4.7). In general, for a given bit-slice, the generation and propagation of carries are mutually exclusive and occur in different cycles. That is, if in the first cycle a carry is generated at bit-slice i, then that carry, if it propagates to the end of the adder and "wraps" around, cannot propagate beyond bit-slice i. This mutual exclusivity of G_i and P_i signals may be similarly applied to all the other bit-slices and combining this with the elimination of C_{-1} (by substituting its definition) leads to a residue adder that has the same delay (as measured in terms of gates) as the adder of Figure 4.6 but which requires only a little more logic. We next show this.

Still assuming the hypothetical two-cycle n-bit adder, the carry-in during the second cycle is the same as the carry-out during the first cycle, and for bit-slice 0, we may express this as

$$\begin{aligned}
C_0^2 &= G_0 + \mathbf{P_0 C_{-1}^2} \\
&= G_0 + P_0 C_{n-1}^1 \\
&= G_0 + \mathbf{P_0}(\mathbf{G_{n-1} + P_{n-1} G_{n-2} + \cdots + P_{n-1} P_{n-2} P_{n-3} \cdots P_1 G_0}) \\
&= G_0 + \mathbf{G_{n-1} P_0 + P_{n-1} G_{n-2} P_0 + \cdots + P_{n-1} P_{n-2} P_{n-3} \cdots P_2 G_1 P_0}
\end{aligned}$$

In each of these expressions, we have split the defining expression for each carry into two parts: the part in normal font corresponds to the usual carry generate-propagate, and the part in bold font corresponds to a carry that (in the first cycle) propagates to the end of the adder and eventually becomes (in the second cycle) an end-around carry that propagates back again. Proceeding as above, for bit-slice 1 we determine that there is a carry if one is generated in that position (during the first cycle), or one is generated in the preceding bit position and propagated through (also during the first cycle), or the carry into the adder is propagated through (during the second cycle).

Thus we have

$$\begin{aligned}
C_1^2 &= G_1 + P_1 G_0 + \mathbf{P_1 P_0 C_{-1}} \\
&= G_1 + P_1 G_0 + \mathbf{P_1 P_0} C_{n-1}^1 \\
&= G_1 + P_1 G_0 + \mathbf{P_1 P_0}(\mathbf{G_{n-1} + P_{n-1} G_{n-2} + \cdots + P_{n-1} P_{n-2} \cdots P_1 G_0}) \\
&= G_1 + P_1 G_0 + \\
&\quad \mathbf{G_{n-1} P_1 P_0 + P_{n-1} G_{n-2} P_1 P_0 + \cdots + P_{n-1} P_{n-2} \cdots P_2 G_2 P_1 P_0}
\end{aligned}$$

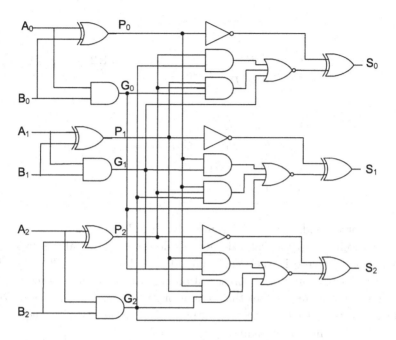

Figure 4.26: Modulo $2^n - 1$ carry-lookahead adder ($n = 3$)

Repeating this process for all the other carries, we arrive at a general formulation for C_i, where $0 \leq i \leq n-1$:

$$\begin{aligned}
C_i^2 &= G_i + P_i G_{i-1} + P_i P_{i-1} G_{i-2} + \cdots P_i P_{i-1} \cdots P_1 G_0 + \mathbf{P_i P_{i-1} \cdots P_0} C_{-1}^2 \\
&= G_i + P_i G_{i-1} + P_i P_{i-1} G_{i-2} + \cdots P_i P_{i-1} \cdots P_1 G_0 + \mathbf{P_i P_{i-1} \cdots P_0} C_{n-1}^1 \\
&= G_i + P_i G_{i-1} + P_i P_{i-1} G_{i-2} + \cdots \mathbf{P_i P_{i-1} \cdots P_1 G_0 P_i P_{i-1} \cdots P_0 (G_{n-1}} \\
&\quad +\mathbf{P_{n-1} G_{n-2}} + \cdots + \mathbf{P_{n-1} P_{n-2} P_{n-3} \cdots P_1 G_0}) \\
&= G_i + P_i G_{i-1} + P_i P_{i-1} G_{i-2} + \cdots P_i P_{i-1} \cdots P_1 G_0 \mathbf{G_{n-1} P_i P_{i-1} \cdots P_0} \\
&\quad +\mathbf{P_{n-1} G_{n-2} P_i P_{i-1} \cdots P_0} + \cdots + \mathbf{G_{i+1} P_i P_{i-1} \cdots P_0 P_{n-1} \cdots P_{i+2}}
\end{aligned}$$

(The superscripts appear on only one side of each final equation, and, as they have served their purpose, we will henceforth drop them.)

As an example the logic equations for a modulo-7 adder ($n = 3$) with two representations for zero are

$$G_i = A_i B_i \qquad i = 0, 1, 2$$
$$P_i = A_i \oplus B_i$$
$$C_{-1} = C_2$$
$$C_0 = G_0 + G_2 P_0 + P_2 G_1 P_0$$
$$C_1 = G_1 + P_1 G_0 + G_2 P_1 P_0$$
$$C_2 = G_2 + P_2 G_1 + P_2 P_1 G_0$$
$$S_i = P_i \oplus C_{i-1}$$

and the corresponding logic diagram is shown in Figure 4.26.

The modulo carry-lookahead adder as described so far allows two representations for zero in the result: $00 \cdots 0$ and $11 \cdots 1$ (i.e. $2^n - 1$). To modify it so that only one representation is permissible, it is necessary to detect the latter and ensure that a sum of zero is instead produced: the reader can verify that the correct output will be produced if the equation above for the sum bits is changed from $S_i = P_i \oplus C_{i-1}$ to

$$S_i = (P_i \cdot \overline{P_0^{n-1}}) \oplus C_{i-1}$$

If $P_0^{n-1} = 0$, then this equation reduces to $S_i = P_i \oplus C_{i-1}$, which is the correct formulation for an output other than $11 \cdots 1$. If $P_0^{n-1} = 1$, an output of $11 \cdots 1$ needs to be converted into $00 \ldots 0$. In this case the equation reduces to $S_i = C_{i-1}$. This is evidently correct, since $P_i = 1$, for all i, means that no carry is generated anywhere; and, as then $C_{-1} = 0$, there is also no carry to propagate anywhere. Therefore, for all i, $C_i = 0$ and so $S_i = 0$. The design of the modified residue adder that corresponds to that of Figure 4.26 is shown in Figure 4.27.

The residue carry-lookahead adders just described are both one-level adders, but the designs are readily extensible to other types of lookahead adder, e.g. multilevel, block, etc [5]. Nevertheless, all such designs will suffer from the standard problem of implementing normal carry-lookahead adders in VLSI, namely they lack regularity. For regular, high-speed designs, parallel-prefix adders are probably the best designs. Moreover, these have a structure that is inherently suitable for the increment operation required in modulo-$(2^n - 1)$ addition.

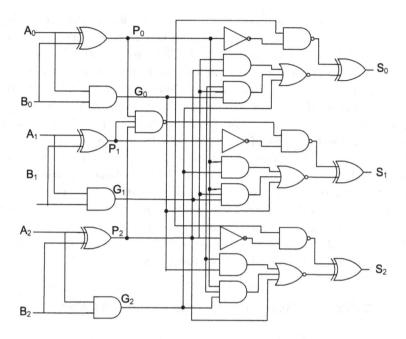

Figure 4.27: Modified modulo $2^n - 1$ carry-lookahead adder ($n = 3$)

4.3.3 Parallel-prefix adder

A seemingly straightforward way to implement modulo-$(2^n - 1)$ adder is to use the structure suggested by Figure 4.19; that is, have one adder that computes $A + B$, one that computes $A + B + 1$, and a multiplexer that selects of these two results. (Appropriate additional logic is required if only one representation of zero is permitted.) In principle, this approach can be used with any design of the underlying adder. But, detailed consideration of the logic required for the two adders will reveal that there is much similarity. Therefore, logic can be shared, and the implicit replication is in fact not necessary in practice. In other words, the two nominal adders can be realized at less than twice the cost of two normal adders. This possibility already exists with certain adder designs, in which, essentially, two intermediate sums that differ by 1 are computed using shared logic. Such adders are therefore "naturally" suitable for modulo-$(2^n - 1)$ addition.

Consider, for example, the conditional-sum adder of Figure 4.11. At each level, two partial sums are computed that differ by unity—one sum,

S^0, under the assumption that there is no carry-in and one, S^1, under the assumption the assumption that there is a carry-in—and one of the two is then selected, at the next level, according to the intermediate carry. This process is repeated until at the last level only one sum, the result, remains. Now, the example of Table 4.1 assumes that the carry into the adder, C_{-1} is 0. If, however, C_{-1} is not known, then then there will be an additional level, with two "final" intermediate sums. (An example is shown in Table 4.4.) One of these can then be selected to be the result when C_{-1} is finally available. A modulo-$(2^n - 1)$ adder that allows two representations for zero can be obtained from the basic conditional-sum adder by using the C_{n-1}^0, the carry-out corresponding to S^0, to select one of the two intermediate sums. Essentially, this sets $C_{-1} = C_{n-1}^0$. And an adder with one representation of zero is obtained by having $C_{-1} = C_{n-1}^0 + P_0^{n-1}$. The latter requires some additional logic, but the composite carry-propagation signal is already available anyway.

Table 4.4: Conditional-sum addition with unknown C_{-1}

i	7		6		5		4		3		2		1		0	
A	1		0		0		1		0		1		0		1	
B	0		0		1		0		1		1		0		1	
	C	S	C	S	C	S	C	S	C	S	C	S	C	S	C	S
C=0	0	1	0	0	0	1	0	1	0	1	1	0	0	0	1	0
C=1	1	0	0	1	1	0	1	0	1	0	1	1	0	1		
C=0	0	1			0	0	1		1	0			0	1		0
C=1	0	1			1	1	0		0	1			1			
C=0	0	1			0		1		1	1			0	1		0
C=1	0	1			1		0		0							
C=0	0	1			1		0		0	0			1			0
C=1	0	1			1		0		0	0			1			1

Since, the general structure of the conditional-sum adder sets it in the class of parallel-prefix adders, it should be possible to easily modify these adders too for modulo-$(2^n - 1)$ addition. This is indeed the case. Consider a normal parallel-prefix adder that has been designed for $C_{-1} = 0$. If we add another level of prefix operators and set $C_{-1} = C_{n-1}$, where the latter carry is from the preceding level (i.e. the last level in a conventional parallel-prefix adder), then we end up with a modulo $2^n - 1$ adder with two permissible representations of zero. As before, a residue adder with just one zero-representation is obtained with $C_{-1} = C_{n-1} + P_0^n$. See Figure 4.28; detailed will be found in [10].

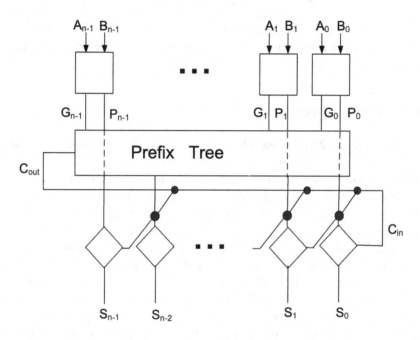

Figure 4.28: Modulo-$(2^n - 1)$ parallel-prefix adder (2 zeros, $n = 3$)

Different parallel-prefix residue adders can be obtained by applying the same basic ideas to any one of the adders in that general class, with the usual tradeoffs between fan-out and delay (prefix-tree depth). There are, however, at least two drawbacks of directly using this approach: the requirement for extra logic (i.e. the extra level of prefix operators) and the high fan-out that may be required of C_{-1}, both of which increase area and operational time. Solutions to these problems can be arrived at by reexamining the design of the residue carry-lookahead adder of Figure 4.26. In the modulo-$(2^n - 1)$ carry-lookahead adder, no extra level of logic is required, because each carry is defined in terms of the carry-propagate and carry-generate signals at all bit positions. If G_i^j and P_i^j are similarly defined—that is, in terms of the G and P bits at all positions—then carries can be recirculated at every level of the prefix tree, and it is not necessary to recirculate a single carry in an additional tree-level [6]. Thus, by defining

$$[G_i^*, P_i^*] = [G_0^i, P_0^i] \bullet [G_{i+1}^{n-1}, P_{i+1}^{n-1}]$$

and $C_i = G_i^*$, the extra prefix operators are eliminated. As an exercise, the reader may wish to try his or her hand in applying this to Slansky and Kogge-Stone adders.

4.4 Addition modulo $2^n + 1$

Addition modulo $2^n + 1$ is considerably more difficult than addition modulo $2^n - 1$, in the sense that it cannot be realized with the same speed or efficiency. The reason for this is apparent if one considers what is required to obtain $|A + B|_m$ from $A + B$; that is, the detection of when $A + B$ exceeds or is equal to the modulus and the correction required to adjust that intermediate result. With $m = 2^n - 1$, both detection and correction are easily implemented: $A+B$ is known to have exceeded the modulus when if there is a carry-out from the addition of A and B; and $A + B$ is known to be equal to the modulus if $A_i \oplus B_i = 1$ for all i. In the former case, the required correction consists of subtracting $2^n - 1$, i.e. subtracting 2^n and adding 1, which is equivalent to ignoring the carry-out and adding a 1 or simply adding the carry-out. And in latter case it consists of adding 1 and ignoring any carry-out. In either case, the correction affects only the least significant end of the adder, in that it consists of the addition of $00 \cdots 01$ or, equivalently, the addition of $00 \cdots 0$ with a carry-in of 1. On the other hand, with $m = 2^n + 1$, two problems arise. First, it is not as easy to determine when the modulus has been exceeded. Second, it is not easy to carry out the required correction. One way around these difficulties, which also extend to multiplication, has been the use of an alternative notation to represent residues modulo $2^n + 1$. Direct addition is then possible but not with the same speed or efficiency as modulo-2^n addition or modulo-$(2^n - 1)$ addition.

4.4.1 Diminished-one addition

Arithmetic operations modulo $2^n + 1$ has frequently been implemented through the use of a different representation —the *diminished-one* representation. In this notation, a number X, where $X > 0$, is represented by the binary equivalent of $X - 1$, which we shall denote by $\widehat{X}$. Zero requires special treatment, e.g. its use may be forbidden or an extra bit may be used in the representations to distinguish between zero and non-zero numbers. We shall assume the latter, and, as $X + 0 = X$, not consider it further in what follows.

With the diminished-one representation, the addition of two numbers, A and B, with diminished-one equivalents $\widehat{A}$ and $\widehat{B}$, is

$$A + B = (\widehat{A} + 1) + (\widehat{B} + 1) = (\widehat{A} + \widehat{B} + 1) + 1$$

If $(\widehat{A} + \widehat{B} + 1) + 1 < 2^n + 1$, then the result is correct in diminished-one form; otherwise $2^n + 1$ must be subtracted to get the correct result. The former condition is equivalent to $(\widehat{A} + \widehat{B} + 1) < 2^n$, which corresponds to the absence of a carry-out, i.e. $C_{n-1} = 0$; while the latter subtraction (which is required when there is a carry-out) is the computation of $(\widehat{A} + \widehat{B} + 1) - 2^n$, which is equivalent to simply ignoring the carry-out. So the diminished-one modulo 2^n+1 addition can be implemented with, say, a parallel-prefix adder in a manner similar to the modulo-$(2^n - 1)$ case but with the end-around carry inverted; that is, with $C_{-1} = \overline{C_{n-1}}$.

The main drawback of the diminished-one representation is that it requires conversions, using adders and subtractors, of the operands and results. A different way in which modulo $2^n + 1$ adders could be designed is to start with the arbitrary-modulus adders of Section 4.1 and then try to optimize the logic for the modulus $2^n + 1$. We leave it to the reader to investigate such designs, and we next describe an approach that is superior in most respects.

4.4.2 Direct addition

With the modulus $2^n - 1$, it easy to determine if the intermediate sum, $A + B$, of the two operands, A and B, is equal to or has exceeded the modulus and needs to be corrected to get $A + B \bmod 2^n + 1$. Required correction is easily accomplished by ignoring any carry-out and adding a 1 into the least-significant position of the intermediate result. Since a parallel-prefix adder is well-suited to the concurrent computation of both $A+B$ and $A+B+1$, the correction can easily be done just by including another level of prefix operators [10]. Modulo-$(2^n + 1)$ addition is more difficult: It is not possible to determine directly from the operand bits if $A + B = 2^n + 1$; and there is no single signal (such as a carry-out) that can be used to easily determine if $A + B > 2^n + 1$. Also, directly modifying a parallel-prefix adder, leads to an intermediate result that requires a further subtraction of a 1. But it is not easy to design an adder that concurrently computes $A + B$ and $A + B - 1$ with the same efficiency as one that concurrently computes $A + B$ and $A + B + 1$. Nevertheless it is possible to come close. The basic idea is that if $2^n + 1$ is subtracted to start with and added back if the subtraction left a negative result, then the problem of later having to subtract a 1 is converted into the easier one of having to add a 1. (Note that adding or subtracting 2^n is just setting the most significant bit to 1 or 0.) The details are as follows.

Let X denote the value $A + B - (2^n + 1)$. Then we can distinguish among three cases:

(a) $A + B \geq (2^n + 1)$; that is, $X \geq 0$
(b) $A + B = 2^n$; that is, $X = -1$
(c) $A + B < (2^n + 1)$ and $A + B \neq 2^n$; that is, $X < 0$

In (a), X is evidently the correct result. In (b), the correct result is 2^n, and obtaining that requires the addition back of $2^n + 1$. The addition of 2^n is achieved by setting, S_n, the most significant bit of the result to 1; and the 1 is added as a carry-in, C_{-1}, to an extra level of prefix operator, as in the modulo-$(2^n - 1)$ parallel-prefix adder. And in (c), we nominally need to add back $2^n + 1$, but in this case we observe that, since $A + B < 2^n$,

$$|X + 2^n + 1|_{2^n+1} = |X + 2^n + 1|_{2^n}$$
$$= |X + 1|_{2^n}$$

Therefore, it is sufficient to just add a 1. The result will be in n bits, and S_n, the $(n+1)$st bit of the sum is therefore set to 0. We may summarise all this as

$$|A + B|_{2^n+1} = \begin{cases} X & \text{if } X \geq 0 \\ 2^n + |X + 1|_{2^n} & \text{if } X = -1 \\ |X + 1|_{2^n} & \text{otherwise} \end{cases} \quad (4.11)$$

Equation 4.11 may be implemented as follows. A carry-save adder (CSA) is used to reduce the three operands, A, B, and $-(2^n + 1)$ to two: a partial-carry, $\widetilde{C}$, and a partial-sum, $\widetilde{S}$. (We assume two's-complement representation for negative numbers.) $\widehat{C}$ and $\widehat{S}$ are then assimilated in an n-bit parallel-prefix adder with an extra level of prefix operators to absorb any 1 that needs to be added during the "correction" phase. All that remains, then, is how to distinguish among the three cases above.

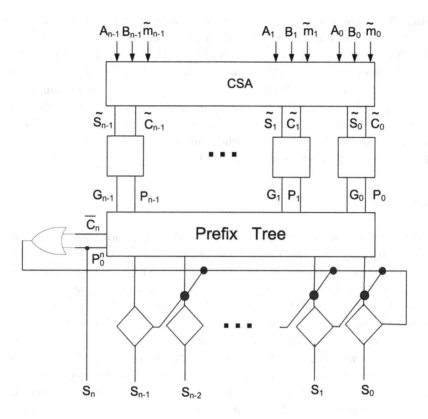

Figure 4.29: Modulo $2^n + 1$ parallel-prefix adder

The bits A_n and B_n will be 0s, and bit n (the most significant bit) of the representation of $-(2^n + 1)$ will be 1. Therefore, bit n of the result of the assimilation will be 1—that is, the result would be negative— only if C_{n-1}, and therefore C_n, is 0. Thus in case (c) detected, and the correction is by adding the inverse of the end-around-carry; that is, $C_{-1} = \overline{C_n}$, and S_n is set to 0. And in case (b), the result of the assimilation would be $111\cdots 1$. But this result occurs only if all the propagate signals $P_i \stackrel{\triangle}{=} \widetilde{C}_i \oplus \widetilde{S}_i$ are 1s; that is if $P_0^n \stackrel{\triangle}{=} P_n P_{n-1} \cdots P_0 = 1$. This suffices to detect case (b). The necessary "correction" here is accomplished by adding P_o^n as an end-around-carry and setting $S_n = P_0^n$.

An appropriate architecture is shown in Figure 4.29. Note that whereas a CSA would normally consists of a sequence of full adders, here some simplification is possible, since m, the modulus, is known: its representation is $1011\cdots 1$. With a normal CSA, the logic equations for the full adder at bit-slice i are $\widetilde{S}_i = (A_i \oplus B_i) \oplus m$ and $\widetilde{C}_i = A_i B_i + m(A_i \oplus B_i)$. So for the bit-slices where $m_i = 1$, $\widetilde{S}_i$ may be simplified to $\overline{A_i \oplus B_i}$; and where $m_i = 0$, it may be simplified to $A_i \oplus B_i$. Similarly $\widetilde{C}_i$ may be simplified to $A_i B_i + (A_i \oplus B_i)$, when $m_i = 1$, or to $A_i B_i$, when $m_i = 0$. A different formulation of Equation 4.11 can be combined with a redefinition of • to yield a modulo-$(2^n + 1)$ adder without an extra level of prefix operators; the reader will find relevant details in [20].

4.5 Summary

Adders for residue arithmetic are derived from adders for conventional binary arithmetic, for the very simple reason that at the end of the day residue digits are represented in conventional binary. Therefore, all of the usual techniques used in the design of conventional adders are applicable here. In essence, modulo-m addition (with operands $0 \leq A, B < m$) requires the production of two intermediate results, $A + B$ and $A + B - m$, one of which is selected as the final result. Regardless of the addition technique used, two distinct adders are not strictly required: some logic may be shared between the two nominal adders. In the case of the special moduli, $2^n - 1$ and $2^n + 1$, such sharing requires very little extra logic if certain adders are used, those adders being ones that would "naturally" produce the two intermediate results (or equivalents).

References

(1) A. Beaumont-Smith and C.-C. Lim, 2001. "Parallel prefix adder design". In: *Proceedings, 15th International Symposium on Computer Arithmetic*.

(2) R. P. Brent and H. T. Kung, 1982. A regular layout for parallel adders. *IEEE Transactions on Computers*, C-31(3):260–264.

(3) M. A. Bayoumi, G. A. Jullien, and W. C. Miller, 1987. A VLSI implementation of residue adders. *IEEE Transactions on Circuits and Systems*, CAS-34(3):284–287.

(4) R. W. Doran, 1988. Variants on an improved carry-lookahead adder. *IEEE Transactions on Computers*, 37(9):1110–1113.

(5) C. Efstathiou, D. Nikolos, and J. Kalamatianos, 1994. Area-efficient

modulo $2^n - 1$ adder design. *IEEE Transactions on Circuits and Systems-II: Analog and Digital Signal Processing*, 41(7):463–467.

(6) Kalampoukas, L., D. Nikolas, C. Efstathiou, H.T. Vergos, and J. Kalamtianos. 2000. High-speed parallel-prefix modulo $2^n - 1$ adder. *IEEE Transactions on Computers*, 49(7):673–679.

(7) V. Kantabutra, 1993. Designing optimum one-level carry-skip adders. *IEEE Transactions on Computers*, 42(6):759–764.

(8) V. Kantabutra, 1993. Accelerated two-level carry-skip adders – a type of very fast adders. *IEEE Transactions on Computers*, 42(11):1389–1393.

(9) V. Kantabutra, 1993. A recursive carry-lookahead/carry-select hybrid adder. *IEEE Transactions on Computers*, 42(12):1495–1499.

(10) R. Zimmerman. 1999 "Efficient VLSI implementation of modulo $2^n \pm 1$ addition and multiplication". In: *Proceedings, 14th Symposium on Computer Arithmetic*, pp. 158–167.

(11) S. V. Lakshmivaran and S. K. Dhall, 1994. *Parallel Computation Using the Prefix Problem*. Oxford University Press, UK.

(12) A. R. Omondi, 1994. *Computer Arithmetic Systems*. Prentice-Hall, U.K.

(13) B. Parhami. 2000. *Computer Arithmetic*. Oxford University Press, U.K.

(14) J. J. Shedletsky, 1977. Comment on the sequential and indeterminate behaviour of an end-around-carry adder. *IEEE Transactions on Computers*, C-26:271–272.

(15) J. F. Wakerly, 1976. One's complement adder eliminates unwanted zero. *Electronics*:103–105

(16) R. E. Ladner and M. J. Fischer. 1980. Parallel prefix computation. *Journal of the ACM*, 27:831–838.

(17) P. M. Kogge and H. S. Stone. 1973. A parallel algorithm for the efficient computation of a general class of recurrence relations. *IEEE Transactions on Computers*, 22:786–793.

(18) S. Knowles. 1999. A family of adders. In: *Proceedings, 14th Symposium on Computer Arithmetic*, pp. 30–34.

(19) A. Tyagi. 1993. A reduced area scheme for carry-select adders. *IEEE Transactions on Computers*, 42(10):1163–1170.

(20) C. Efstathiou, H. T. Vergos, and D. Nikolos. 2004. Fast parallel-prefix modulo $2^n + 1$ adders. *IEEE Transactions on Computers*, 53(9):1211–1216.

(21) M. J. Flynn and S. F. Oberman. 2001. *Advanced Computer Arith-*

metic Design. Wiley, New York.
(22) M. Dugdale. 1992. VLSI implementation of residue adders based on binary adders. *IEEE Transactions on Circuits and Systems II: Analog and Digital Signal Processing*, 35(5):325329.
(23) A. A. Hiasat. 2002. High-speed and reduced-area modular adder structures for RNS. *IEEE Transactions on Computers*, 51(1):8489.
(24) D. K. Banaji. 1974. A novel implementation method for addition and subtraction in residue number systems. *IEEE Transactions on Computers*, C-23(1):106108.
(25) S. Piestrak. 1994. Design of residue generators and multioperand modular adders using carry save adders. *IEEE Transactions on Computers*, 42(1):68–77.

Chapter 5

Multiplication

As with addition, multiplication relative to an arbitrary modulus is rather difficult. Modular multiplication may be implemented in the form of table-lookup, combinational logic, or a mixture of combinational logic and table-lookup. Pure table-lookup is rarely implemented, except, perhaps, for very small moduli; for large moduli, the tables required will be large and slow. Table-lookup in conjunction with a small amount of combinational logic also has few advantages when an arbitrary modulus is used, although it is an improvement on pure table-lookup. More efficient combinational-logic multiplier-designs are possible when the modulus is one of the special ones: $2^n, 2^n - 1$ and $2^n + 1$.

With arbitrary moduli, there are two main ways to proceed. Let A and B be the two operands whose modular product is to be computed. Then the first method is to proceed directly from the definition of a residue—that is, as the remainder from an integer division—and first compute the product AB and then reduce that relative to the given modulus. And the second is to compute AB but carry out the modular reduction during that multiplication: multiplication is essentially a series of additions, and the idea is to modular-reduce each operand and then add. We shall therefore start with a review of algorithms for conventional multiplication and division.

With the first general approach above, a straightforward way to compute the modular product, $|AB|_m$, through division is to simply multiply A and B, divide the result by m, and then take the remainder. The key problem, then, is how to efficiently divide two numbers. The division here is conventional division, for which there are two main classes of algorithms. The first class consists of *subtractive algorithms*, which are variants of the standard paper-and-pencil division algorithms. In these algorithms, two sequences are computed such that one converges to the quotient and the other converges to the remainder. The other class of division algorithms consists of *multiplicative algorithms*, in which two sequences are computed (as continued products) such that as the values of one sequence converge to some constant (usually zero or unity), the values of the other converge to the quotient. This latter class therefore requires that the remainder be obtained through additional computations— typically, another multiplication and a subtraction (which is just an addition)— since it is not readily available. In practice, for modular multiplication actual division is usually not carried out, but the essential aspects of division algorithms will still be employed.

This chapter has five main sections. The first two sections are a review of the main algorithms used for multiplication and division in conventional number systems; the reader who is familiar with these algorithms may skip these sections. The third section shows how these algorithms may be employed as the basis of algorithms for modular multiplication, with respect to an arbitrary modulus, and also discusses implementations that use pure table-loookup or a combination of that with some combinational logic. The fourth and fifth sections deal with multiplication algorithms, and corresponding implementations, for the special moduli, $2^n - 1$ and $2^n + 1$. (Multiplication modulo-2^n is, of course, just conventional binary multiplication.)

Throughout what follows, "multiplication" and "division" without the qualifier "modular" (or "residue") will mean conventional multiplication and division. It should also be noted that because residue multiplication is digit-wise parallel, in what follows, for operands, "multiplicand" and "multiplier" will be residues.

5.1 Conventional multiplication

Almost all algorithms for the multiplication of unsigned binary numbers are basically versions of the standard procedure for paper-and-pencil multiplication of decimal numbers. The latter process consists of the following steps. Starting from the rightmost digit, each digit of the multiplier is multiplied by the multiplicand. The product of each such "sub-multiplication" is shifted left by i places, where i is the position of the corresponding digit of the multiplier (counting from right to left and starting at zero). The shifts produce *partial products* that are then added up to obtain the complete product.

Multiplication

```
                              5   =   00101        multiplicand
                              11  =   01011        multiplier
                                      00000        Initial partial product
5  =     00101    multiplicand         00101        add 1st multiple
11 =     01011    multiplier           00101
         00000    Initial partial product
         00101    add 1st multiple    000101        shift right
         00101                         00101        add 2nd multiple
         00101    add 2nd multiple    001111
         001111
         00000    add 3rd multiple    0001111       shift right
         0001111                       00000        add 3rd multiple
         00101    add 4th multiple    0001111
         00110111
         00000    add 5th multiple    00001111      shift right
         55  =    final produce        00101        add 4th multiple
         000110111                     00110111

                                      000110111    shift right
                                      00000        add 5th multiple
                                      000110111    final produce
```

Figure 5.1: Sequential multiplication

5.1.1 Basic binary multiplication

In basic binary multiplication, the process is similar, but the formation of a partial product is simplified by the fact that the only possible values for a digit are 0 and 1; so each multiplicand-multiple is either just the multiplicand shifted or zero. The simplest possible implementation of a multiplier, a *sequential multiplier*, follows a roughly similar procedure but with two main differences. First, the partial products are formed one at a time and added to a running total that ends up as the final product. Second, instead of shifting each multiplicand-multiple to the left, to form a partial product, an equivalent effect is achieved by instead shifting the accumulated value to the right, by one place for each partial product. (Note that this means that a partial product now is just a multiplicand-multiple.) An example is shown in Figure 5.1. If one views the sequential multiplier as a starting point, then the speed of multiplication can be improved by either performing the additions at a faster rate or by reducing the number of partial products to be added.

```
   7 =   00111   multiplicand
  15 =   01111   multiplier
```

00000	initial partial sum
00000	initial partial carry
<u>00111</u>	add 1st multiple
00111	1st partial sum
00000	1st partial carry
000111	shift partial sum right
00111	add 2nd multiple
<u>00000</u>	add 1st partial carry
001001	2nd partial sum
00011	2nd partial carry
0001001	shift partial sum right
00111	add 3rd multiple
<u>00011</u>	add 2nd partial carry
0011001	3rd partial sum
00011	3rd partial carry
000111001	shift partial sum right
00000	add 4th multiple
<u>00011</u>	add 3rd partial carry
000001001	4th partial sum
00011	4th partial carry
000111001	shift partial sum right
00000	add 5th multiple
<u>00011</u>	add 4th partial carry
000001001	5th partial sum
00011	5th partial carry
0000001001	shift partial sum right
<u>00011</u>	propagate final carries
105 = 001101001	

Figure 5.2: Sequential carry-save multiplication

The addition of a partial product can be speeded up by observing that carry-propagation, the dominant factor in addition, is not necessary with every addition in a sequence of additions. Instead, carries from one addition may be saved and added, with appropriate significance shifts, to the operand of the succeeding addition; only in the last addition do the carries need to be propagated. Figure 5.2 demonstrates this, in an example that corresponds to Figure 5.1. Except for the last addition, which requires an adder of the types described in Section 4.1, all additions are carried out in a *carry-save adder* (CSA), which consists of just a set of full adders.

In this context, a full adder is also known as a *3:2 counter*[1], as it reduces three inputs to two outputs. The outputs of a carry-save adder are *partial-sum* and *partial-carry* bits that must be *assimilated* (i.e. added, with the carry-bits shifted, in a carry-propagate adder) to yield a single output in conventional form. The corresponding binary multiplier has the form shown in Figure 5.3. (We assume an appropriate register to separate the inputs of the CSA from the outputs.)

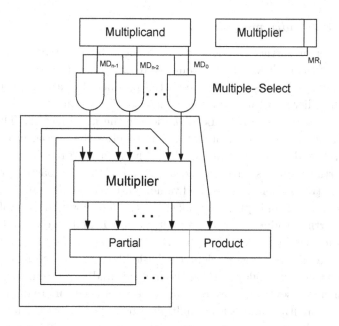

Figure 5.3: Sequential binary (radix-2) multiplier

---

[1]Specialized counters that are designed for regular layouts of tree topologies are also known as *compressors*.

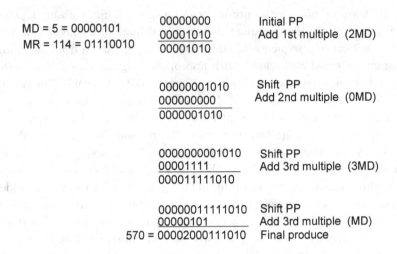

Figure 5.4: Radix-4 multiplication

5.1.2 High-radix multiplication

The number of partial products can be reduced by, effectively, increasing the radix of the multiplication—from two to, say, four or eight. A straightforward way to do this is form the partial products by taking several bits of the multiplier at a time: two for radix-4 multiplication, three for radix-8 multiplication, and so forth. Let MD denote the multiplicand. Then, directly taking two multiplier-bits at a time requires the partial products 0, MD, $2MD$, and $3MD$; and taking three bits at a time requires the additional partial products $4MD$, $5MD$, $6MD$ and $7MD$. The partial products (or, alternatively, their accumulated value) must at each step now be shifted by the number of multiplier-bits taken. Figure 5.4 shows an example.

The partial products that are a multiple of MD and a power of two can easily be formed by shifts of MD; but the others may require lengthy carry-propagate additions. Nevertheless, the basic idea can be usefully extended to larger radices, as follows. Both the multiplicand and the multiplier are partitioned into several pieces, each piece being of one or more digits. Each piece of the multiplicand is then multiplied by each piece of the multiplier, and the results are then added up, with appropriate shifting for significance. Figure 5.5 shows an example-the multiplication of 1234 and 5678. Figures 5.6 shows a corresponding for a 16-bit×16-bit multiplier. In implementation, using just combinational logic for such a multiplier is not very effective: a carry-propagate adder is required for each "small" multiplication, and all the adders in the tree are nominally carry-propagate adders; this increases both cost and operational time. On the other hand, the full use here of carry-save adders leads to a structure that is similar to others discussed below but inferior in several aspects. A better approach is to use look-up tables (ROMs) to form the partial products—that is, replace the small multipliers of Figure 5.6 with ROMs—and then add them up in a tree of carry-propagate adders or a tree of carry-save adders with an assimilating carry-propagate adder. And one can readily envisage an all-ROM implementation. Evidently there is a trade-off between the ROM sizes and the depth of the tree. The main advantage of these multiplier designs is that where it is preferable to use (mostly) table-lookup, they will result in smaller ROMs than would be the case with direct ROM-multiplication. Nevertheless, such multipliers have not been commonly implemented for conventional multiplication—for the simple reason that there is a generally better way (Booth's Algorithm) to accomplish high-radix multiplication. A partitioned-operand modular multiplier has been proposed for modulo-$(2^n - 1)$ multiplication [2] and is briefly described in Section 5.4, but even for that limited case it is of dubious worth.

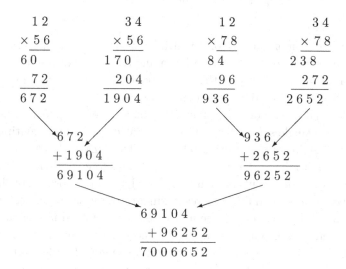

Figure 5.5: High-radix multiplication by operand-splitting

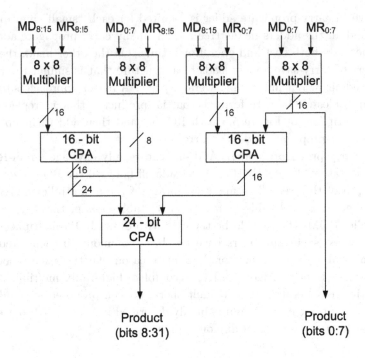

Figure 5.6: Split-operand multiplier

The problem of hard-to-form multiplicand-multiples in high-radix multiplication can be partially solved by taking advantage of the fact that the addition of several multiples that correspond to a string of 0s in the multiplier may be replaced with a single addition of zero, and that of adding multiples that correspond to a string of 1s may be replaced with a single addition and a single subtraction. To see the latter, consider a multiplication of MD by a string of k 1s. Since $MD \times 111 \cdots 1 = MD(2^{k-1} + 2^{k-2} + \cdots 2^0) = MD(2^k - 1)$, it follows that instead of k additions (one for each 1), it is sufficient to have a single subtraction corresponding to the position of the least significant 1 and an addition corresponding to the position immediately after the most significant 1. If the multiplier is partitioned into equal-length pieces, then for a given piece, it will be necessary to determine whether it is at the start of a string of 1s, the end of a string of 1s, the start and end of a string of 1s, and so forth. For each partition, this may be accomplished by examining the last bit of the preceding partition. Thus for radix-4 and radix-8 multiplication the partial products are formed according to the rules in Tables 5.1 and 5.2. Essentially, what these rules capture is an "on-the-fly" *recoding* of the multiplier, from conventional binary representation to the redundant-signed-digit notation of Chapter 1: the digit-set for the radix-4 is $\{\bar{2}, \bar{1}, 0, 1, 2\}$, and that for the radix-8 is $\{\bar{3}, \bar{2}, \bar{1}, 0, 1, 2, 3\}$. A corresponding example multiplications are shown in Figure 5.7; it is evident that the number of hard-to-form multiples has been reduced. The multiplication algorithm based on these rules is known as *Booth's Algorithm*, and it is worth noting that it is useful even with radix-2, because it simplifies the multiplication of signed numbers.

Table 5.1: Radix-4 Booth Algorithm

$MR_{i+1,i}$	MR_{i-1}	Action
00	0	Shift PP 2 places
00	1	Add MD; shift PP 2 places
01	0	Add MD; shift PP 2 places
01	1	Add 2×MD; shift PP 2 places
10	0	Subtract 2×MD; shift PP 2 places
10	1	Subtract MD; shift PP 2 places
11	0	Subtract MD; shift PP 2 places
11	1	Shift PP 2 places

Given the relative decreases that have occurred in the costs of hardware, most implementations of multipliers are no longer sequential (Figure 5.3). Instead of a single carry-save adder, the number and duration of required addition cycles are reduced by "unrolling" the multiplier "loop", one or more times, so that two or more carry-save adders are used to add several partial products in a single cycle. Typically, in current multipliers, complete "unrolling" is done so that the final design consists of a set of carry-save adders with a carry-propagate adder at the end. Figure 5.9 shows an example for a 5-bit×5-bit multiplication. Booth recoding may also be used in such a multiplier to reduce the number of partial products to be added.

In the multiplier of Figure 5.9, the (logical) delay through the tree of carry-save adders is proportional to the number of partial products. This may be improved upon by adding as many partial products as possible at each level of the tree; that is, by using more than one carry-save adder at each level.

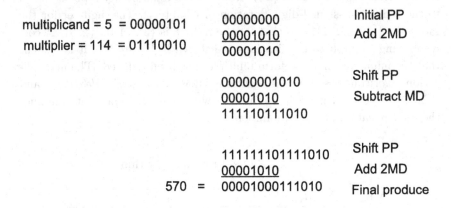

Figure 5.8: Multiplication with radix-8 Booth Algorithm

Since a carry-save adder is a 3-input/2-output unit, partial products are taken in as many groups of three as are possible, with a carry-save adder for each group. Thus, for unrecoded n-bit×n-bit multiplication, the first level of the addition tree consists of about $n/3$ carry-save adders. At the second level, the outputs of these adders, plus any leftovers from the first level, are again grouped into threes, with appropriate significance shifting. This process is repeated until there are only one partial-sum and one partial-carry left, and these are then assimilated in a carry-propagate adder. An example is shown in Figure 5.10, for 5-bit×5-bit multiplication. This type of multiplier is known as a *Wallace-tree multiplier*, and the logical delay through its reduction tree is proportional to $\log_{3/2} n$. Although the operational time here appears to be better than that of the simple parallel-array multiplier, the Wallace-tree has a less regular structure, and a compact layout is therefore more difficult. Therefore, in current VLSI technology, the area, interconnection-delays, etc., from this design can be larger than is apparent from Figure 5.10. Consequently, the actual operational time may not be substantially different from that of the parallel-array. As with the parallel-array multiplier, Booth recoding may be used here to reduce the number of partial products and, therefore, the depth of the tree. In addition to the simple parallel-array and Wallace-tree multipliers, there are several other arrangements that are possible when the multiplier "loop" is completely unrolled. For these, we leave it to the reader to refer to the published literature [5, 6, 12].

Table 5.2: Radix-8 Booth Algorithm

$MR_{i+2,i+1,i}$	MR_{i-1}	Action	
000	0	Shift PP 3 places	
000	1	Add MD; shift PP 3 places	[0, 0, 1]
001	0	Add MD; shift PP 3 places	[0, 2, −1]
001	1	Add 2×MD; shift PP 3 places	[0, 2, 0]
010	0	Add 2×MD; shift PP 3 places	[4, −2, 0]
010	1	Add 3×MD; shift PP 3 places	[4, −2, 1]
011	0	Add 3×MD; shift PP 3 places	[4, 0, −1]
011	1	Add 4×MD; shift PP 3 places	[4, 0, 0]
100	0	Subtract 4×MD; shift PP 3 places	[−4, 0, 0]
100	1	Subtract 3×MD; shift PP 3 places	[−4, 0, 1]
101	0	Subtract 3×MD; shift PP 3 places	[−4, 2, −1]
101	1	Subtract 2×MD; shift PP 3 places	[−4, 2, 0]
110	0	Subtract 2×MD; shift PP 3 places	[0, −2, 0]
110	1	Subtract MD; shift PP 3 places	[0, −2, 1]
111	0	Subtract MD; shift PP 3 places	[0, 0, −1]
111	1	Shift PP 3 places	

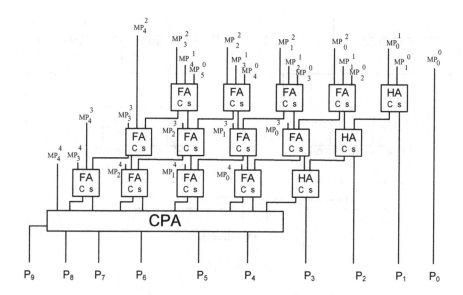

Figure 5.9: Parallel-array multiplier ($n = 5$)

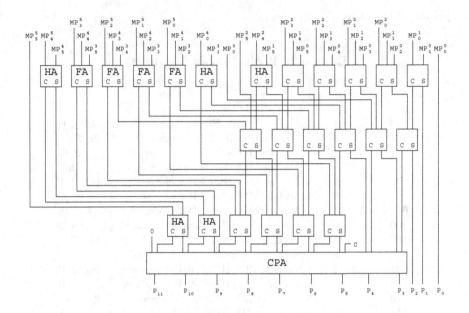

Figure 5.10: Wallace-tree multiplier ($n = 5$)

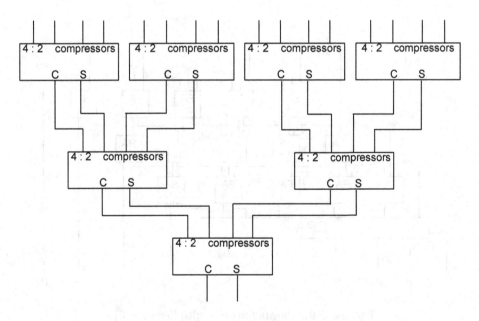

Figure 5.11: Multiplier of 4:2 compressors

So far, we have assumed that the basic element in a carry-save adder is a 3:2 counter (i.e. a full adder). Up to a point, faster multipliers can be obtained by increasing the degree of compression. Thus, for example, 5:3 and 7:3 counter have been used. As indicated above, for regular topologies, it is common to use compressors, which are specialized forms of counters. A (p,q) compressor has p inputs and q outputs; but the numbers reflect the fact that the carry inputs and outputs are from and to adjacent compressors (on the same level) rather than to compressors on the next level. So a 4:2 compressor is just an instance of a 5:3 counter. As an example, Figure 5.11 shows a 4:2 compressor-tree for the reduction of sixteen operands to two . A straightforward design for a 4:2 compressor is to take two full-adders (i.e. 3:2 compressors) in series. It is, however, possible to design a 4:2 compressor whose delay is less than that of two full adders [10].

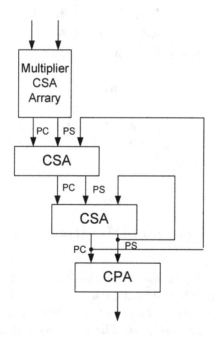

Figure 5.12: Multiply-Accumulate unit

There are many scientific applications (e.g. digital signal processing) in which the accumulation of products (as in the computation of an inner-product) is a fundamental operation that, ideally, should be performed at the highest possible speed. A simple design for an appropriate hardware unit nominally consists of a multiplier and an adder (the accumulator) arranged in series and used repetitively. Implementing this directly produces a structure with two carry-propagate adders, one being the assimilation adder in the multiplier and the other being for the sum-accumulation. Such an implementation can be improved upon by observing that since assimilation is not really necessary until the last addition, the multiplier's carry-propagate adder may be replaced with two carry-save adders. A design for a multiply-accumulate unit is shown in Figure 5.12, of which several slightly different organizations are possible.

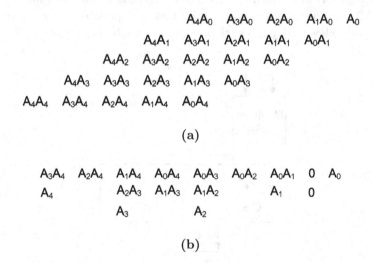

Figure 5.13: Partial-product array in squaring

We conclude this section with a brief mention of a special case in multiplication —squaring. Where squaring is a crucial operation, it may be useful to implement a dedicated hardware unit, rather than carry out the operation in a general-purpose multiplier. The advantage of doing so is that, since the two operands are the same, a specialized unit can be made smaller and faster [8]. Consider, for example, the computation of A^2 by a full multiplication. The array of partial products has the form shown in Figure 5.13(a), for a 5-bit×5-bit multiplication. Every other term in the anti-diagonal has the form A_iA_i, which is equivalent to just A_i, since A_i is 0 or 1. There is also a symmetry around the same diagonal, since $A_iA_j = A_jA_i$. So the two terms

(A_iA_j and A_jA_i) may be replaced with their sum, $2A_iA_j$, which, since multiplication by two is a 1-bit left-shift, is just A_iA_j moved into the next column to the left. Therefore, the matrix of Figure 5.13(a) may be compressed to the equivalent one in Figure 5.13(b). Lastly, consider the terms A_i and A_iA_j occurring in the same column. If $A_i = 0$, then $A_i + A_iA_j = 0$; if $A_i = 1$, and $A_j = 0$, then $A_i + A_iA_j = A_i = A_i\overline{A_j}$; and if $A_i = 1$ and $A_j = 1$, then $A_i + A_iA_j = 2 = 2A_iA_j$. So $A_i + A_iAj = 2A_iA_j + A_i\overline{A_j}$, which corresponds to $A_i\overline{A_j}$ in the same column and $2A_iA_j$ moved into the next column to the left. This may be used to further reduce the number of terms in columns of large arrays; we leave it as an exercise for the reader to try it with, for example, an 8-bit A.

5.2 Conventional division

Multiplication consists of the addition of partial products that are shifted multiples of the multiplicand and each digit of the multiplier. Conversely, direct division consists of subtractions (i.e. additions of negations) of the shifted multiples of the divisor, from a *partial remainder* that is initially equal to the dividend and finally less than the divisor. The multiple that is subtracted at each step is the product of the divisor and the quotient-digit determined at that step. This process is *subtractive division*. An alternative approach is *multiplicative division*, in which the basic operation is multiplication rather than subtraction. One popular algorithm of the latter type consists, essentially, of a computation of the reciprocal of the divisor, which reciprocal is then multiplied by the dividend to obtain the quotient. There are a variety of algorithms for both subtractive and multiplicative division, but in what follows we shall discuss only three of these. For others, the reader should consult texts devoted to general aspects of computer arithmetic [5, 6, 12].

5.2.1 Subtractive division

Direct binary division may be realized as follows. Let N be the dividend, D be the divisor, $Q^{(i)}$ be the (partial) quotient at the end of the i-th step, and $R^{(i)}$ be the corresponding partial remainder. In general, N and $R^{(i)}$ will be represented in twice as many bits as each of the other operands—$2n$ versus n bits. Initially, $Q^{(0)} = 0$ and $R^{(0)} = N$. At each step, i ($i = 0, 1, 2, \ldots, n-1$), $R^{(i+1)}$ is formed by the subtraction $R^{(i)} - D$, if $R^{(i)}$ is positive, or the addition $R^{(i)} + D$, if $R^{(i)}$ is negative; $Q^{(i+1)}$ is incremented[2] by 1 if in the preceding step a subtraction took place; and the partial remainder and the partial quotient are then shifted left by one bit (i.e. effectively multiplied by two). After n steps, $Q^{(n)}$ will be the quotient and $R^{(n)}$, if it is positive, will be the remainder; if $R^{(n)}$ is negative, then the remainder is $R^{(n)} + D$. This procedure is known as *non-restoring division*, and it is essentially a derivation of paper-and-pencil division.[3] We may represent the procedure with the recurrences $(i = 0, 1, 2, \ldots, n-1)$

$$Q^{(0)} = 0$$
$$R^{(0)} = N$$
$$R^{(i+1)} = \begin{cases} 2R^{(i)} - D & \text{if } 2R^{(i)} \geq 0 \\ 2R^{(i)} + D & \text{if } 2R^{(i)} < 0 \end{cases}$$
$$q_i = \begin{cases} 1 & \text{if } 2R^{(i)} - D \geq 0 \\ 0 & \text{if } 2R^{(i)} - D < 0 \end{cases}$$
$$Q^{(i+1)} = 2Q^{(i)} + q_i$$
$$R = \begin{cases} R^{(n)} & \text{if } R^{(n)} \geq 0 \\ R^{(n)} + D & \text{otherwise} \end{cases}$$

A direct, implementation of this procedure would be rather slow because each step requires a complete carry-propagate addition. (This is what makes division fundamentally a more difficult operation than multiplication.) In multiplication, carry-propagations at each step can be avoided by using carry-save adders and postponing all carry-propagation to the last step. This is not entirely possible in division, because of the need to know the sign of each partial remainder, but good use can still be made of the basic carry-save idea.

Consider a modification of non-restoring division in which the additions and subtractions are carried out in carry-save adders, with final assimilation after the last step. The partial remainder will not be fully known at each step, since it will be in partial-carry/partial-sum form, but an approximation of it can be obtained by assimilating a few of the leading partial-carry/partial-sum bits. This approximation is then compared with

---

[2] This is not a real addition: either a 1 or a 0 is appended at the end of the partial quotient.

[3] In *restoring division*, before the next step is started, the partial remainder is completely restored to its old value (by adding back the divisor) if a subtraction left it negative. This is therefore a less efficient process, although it more closely mirrors paper-and-pencil division, in which the restoration is done "on the side".

the divisor, and the next bit of the quotient is selected. Because only an approximation is used, the result will sometimes be incorrect. To handle this, a redundant-signed-digit set (RSD) is now used for the quotient.[4] Suppose, for example, that the selected quotient digit is 1 when it should be 0 and the next digit should be 1. In such a case, in the next step, the quotient digit is then selected to be $\bar{1}$ instead of 1. The end result is then correct, as both $1\bar{1}$ and 01 represent the same number. In short, by using RSD notation, errors made at some step can be corrected in subsequent steps. The algorithm obtained by the modifications just described is known as *SRT division*, and it is quite popular, because its implementations have good cost:performance ratios. Based on the above, the basic binary SRT algorithm may be represented by the recurrences

$$q_i = \begin{cases} 1 & \text{if } 2R^{(i)} > 0 \\ 0 & \text{if } -D < 2R^{(i)} < D \\ \bar{1} & \text{if } 2R^{(i)} < 0 \end{cases}$$

$$R^{(i+1)} = \begin{cases} 2R^{(i)} - D & \text{if } 2R^{(i)} > 0 \\ 2R^{(i)} & \text{if } -D < 2R^{(i)} < D \\ 2R^{(i)} + D & \text{if } 2R^{(i)} < 0 \end{cases}$$

In what follows, we shall assume that the operands are in the range $[1/2, 1)$. This is only for convenience and does not affect the actual range of the operands: it is merely a matter of scaling and the assumed positions of the radix-point. The main benefit of forcing operands to be in this range is that the comparisons now change from full-length ones between $2R^{(i)}$ and D or $-D$ to low-precisions ones between $2R^{(i)}$ and a pair of constants.

The study of SRT algorithms usually involves the use of *Robertson diagrams*, which are plots of the next-partial-remainder against the shifted current-partial-remainder. Essentially, such a plot is an expression of the invariant for binary division: $R^{(i+1)} = 2R^{(i)} - q_i D$, where q_i is the quotient digit at step i. Thus the Robertson diagram corresponding to the radix-2 recurrences above is as shown in Figure 5.15. It will be seen that for a given value of $R^{(i)}$, there are two choices for q_i; this is a reflection of the redundancy in the digit-set $\{\bar{1}, 0, 1\}$. The most significant aspect of the redundancy is that it allows some choice in the values against which the partial remainder is compared; they need not be D and $-D$, as an approximation will suffice. Two criteria have commonly been used in the selection of these values: choosing values that increase the probability of q_i

---

[4]See Chapter 1 for a review of such digit-sets.

being 0, which allows for shifting without any arithmetic in that cycle, and values that facilitate simple and efficient comparison. In current dividers, the former is of little value, and it is the latter that is used.

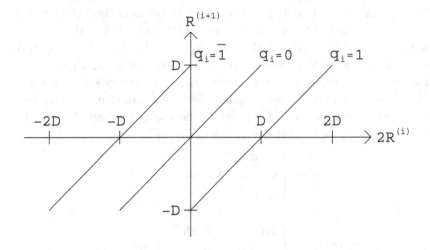

Figure 5.14: Robertson diagram for radix-2 SRT division

As indicated above, the redundancy in the digit-set also means that in the comparisons, the partial remainder, $R^{(i)}$, may be replaced by an approximation, $\widetilde{R}^{(i)}$, that is the same as $R^{(i)}$ in only a few leading bits (with the rest assumed to be 0s). For binary, i.e radix-2, division it can be shown that an approximation that consists of the most significant three bits is sufficient, and that for this suitable comparison constants are $-1/2$ and $1/2$. Thus the core of the radix-2 SRT algorithm is

$$q_i = \begin{cases} 1 & \text{if } 2\widetilde{R}^{(i)} \geq \frac{1}{2} \\ 0 & \text{if } -\frac{1}{2} \leq 2\widetilde{R}^{(i)} < \frac{1}{2} \\ \overline{1} & \text{if } 2\widetilde{R}^{(i)} < -\frac{1}{2} \end{cases}$$

$$R^{(i+1)} = 2R^{(i)} - q_i D$$

In general, the best choice of comparison constants depends on the given value of D. It can be shown that for D in the range $[1/2, 1)$, the optimal value has to be selected from five different ones, according to the sub-range in which D lies. The SRT algorithm produces a quotient that is in RSD form and which therefore requires a conversion. This can be done "on-the-fly", i.e. as the quotient digits are produced [7]. The remainder also needs a conversion as it will be in partial-sum/partial carry form. As in multiplication, this assimilation is done in a carry-propagate adder. An example of a radix-2 SRT division is shown in Figure 5.15.

dividend = 0.000110 = 3/32   divisor = 0.1101 = 13/16

partial remainder	partial quotient	
0.000110	0 . - - - - - - - - - 0	shift
0.001100	0 . - - - - - - - - 0 0	shift
0.011000	0 . - - - - - - 0 0 0	shift
0.110000 1.0011 1.111100	0 . - - - - - - 0 0 0 1	subtract
1.111000	0 . - - - - - 0 0 0 1 0	shift
1.110000	0 . - - - - 0 0 0 1 0 0	shift
1.100000	0 . - - - 0 0 0 1 0 0 0	shift
1.000000 0.1101 1.110100	0 . - - - - 0 0 0 1 0 0 0 1	add
1.101000	0 . - - 0 0 0 1 0 0 0 1 0	shift
1.010000	0 . 0 0 0 1 0 0 0 1 0 1	shift
1.000100	0 . 0 0 1 0 0 0 1 0 1 0	shift

Quotient = $\dfrac{1}{8} - \dfrac{1}{128} - \dfrac{1}{512} \approx \dfrac{3}{26}$

Figure 5.15: A radix-2 SRT division

In multiplication, the use of Booth recoding reduces the number of difficult-to-form multiples of the multiplicand and so makes it easy to design multipliers with radices larger than two. Now, what Booth's Algorithm does is, essentially, an "on-the-fly" recoding of the multiplier, from conventional form to RSD form. If we view division as the inverse of multiplication—the quotient corresponds to the multiplier, and the subtractions in division correspond to the additions in multiplication —then we should expect that recoding the quotient will have a similarly beneficial effect. This is indeed the case, and on this basis may be developed a variety of high-radix SRT algorithms. For example, the core of a radix-4 SRT algorithm with the digit-set $\{\overline{2}, \overline{1}, 0, 1, 2\}$ is

$$q_i = \begin{cases} 2 \text{ if } \frac{3D}{2} \leq 4R^{(i)} \\ 1 \text{ if } \frac{D}{2} \leq 4R^{(i)} < \frac{3D}{2} \\ 0 \text{ if } -\frac{D}{2} < 4R^{(i)} < \frac{D}{2} \\ \overline{1} \text{ if } -\frac{3D}{2} < 4R^{(i)} \leq -\frac{D}{2} \\ \overline{2} \text{ if } -\frac{3D}{2} \leq 4R^{(i)} \end{cases}$$

$$R^{(i+1)} = 4R^{(i)} - q_i D$$

As in the radix-2 case, there exist several choices for the comparison constants, according to the basic Robertson diagram, and both R and D may be replaced with approximations $\widehat{R}$ and $\widehat{D}$. The precision required for the comparisons depends on the degree of redundancy in the digit-set. For the minimally-redundant radix-4 digit-set, i.e. $\{\overline{2}, \overline{1}, 0, 1, 2\}$, it can be shown that the four most significant bits each of the divisor and the partial remainder are sufficient; and for the maximally-redundant digit-set, i.e. $\{\overline{3}, \overline{2}, \overline{1}, 0, 1, 2, 3\}$, two bits of the partial remainder and three of the divisor are required. Extending SRT division to radices higher than four is difficult. The minimally-redundant radix-8 is barely worthwhile, but the maximally-redundant radix-8 is not. The reasons for such difficulties are the same as those in high-radix multiplication: multiples become more difficult to form as the radix increases. The minimally-redundant radix-8 requires the multiple $3D$, but the maximally-redundant radix-8 requires the multiples $3D, 5D$, and $7D$. It is, however, possible to easily implement what are essentially very-high-radix dividers by overlapping two lower-radix dividers.

Thus, for example, a radix-16 divider may be obtained by overlapping two radix-4 dividers.

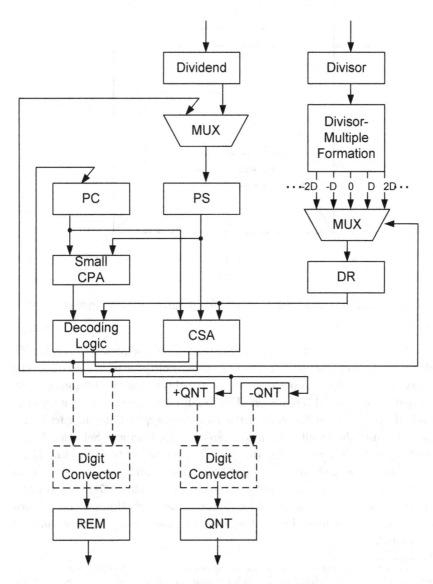

Figure 5.16: Sequential SRT divider

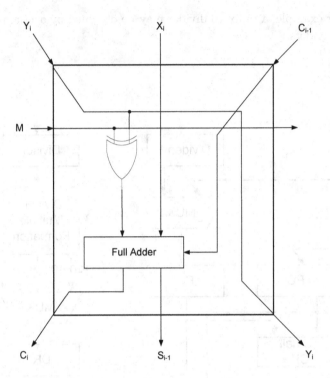

Figure 5.17: Add/Subtract cell for parallel-array divider (radix-2)

Figure 5.16 shows the general organization of a sequential radix-2 SRT divider. (This corresponds to the sequential multiplier of Figure 5.3.) The carry-save adder (CSA) performs the main subtraction or addition required to reduce the partial remainder. The Small CPA (carry-propagate adder) assimilates a few leading bits of the partial-carry/partial-sum current remainder, and the result is then compared, in the Quotient-Selection Logic, with a few leading bits of the divisor.[5] The output of the Decoding Logic is the next quotient digit, in signed-digit form, and this is immediately converted (using the "on-the-fly" conversion technique [7]) into conventional notation. For higher-radix division, more multiples of the divisor are required, and more bits of the partial remainder and divisor must be compared.

---

[5]Typical implementations have replaced the Small CPA and Quotient-Selection Logic with a PLA, or similar structure, in which the assimilation and comparison are combined and "built-in".

Multiplication

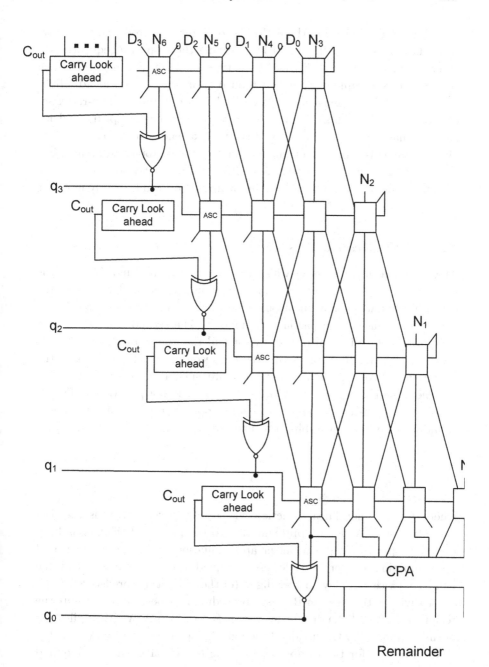

Figure 5.18: Parallel-array divider (radix-2)

"Unrolling" the loop in the multiplier of Figure 5.3 produced the array-multiplier of Figure 5.10. Similarly, unrolling the loop in a sequential divider produces the array-divider of the type shown in Figures 5.17 (basic cell) and 5.18 (complete array), for 6-bit divisor and 3-bit dividend. High-radix array dividers may be obtained by likewise unrolling high-radix sequential dividers. Array-dividers have very rarely been implemented for conventional Arithmetic—the use of carry-lookahead is such that each row of the divider is almost equivalent to a full carry-propagate adder—but they may have some use in residue multiplication (relative to an arbitrary modulus), for which just about any implementation will be costly in either time or logic.

5.2.2 Multiplicative division

There are two main ways in which division by repeated multiplication is usually carried out. The first is an algorithm that is similar to the SRT algorithm but with multiplications instead of subtractions as the primary means for reducing the partial remainder.[6] The second consists of first computing the reciprocal of the divisor and then multiplying that by the dividend. We shall here briefly discuss the second method; for the first, the reader should consult the relevant literature, such as [5, 6, 12].

The basic algorithm for division-by-reciprocal-multiplication is derived from the Newton-Raphson procedure for computing the root of a non-linear equation. The Newton-Raphson recurrence is

$$X_{i+1} = X_i - \frac{f(X_i)}{f'(X_i)}$$

where f is the non-linear function whose root is sought, f' is the first derivative of f, X_i is an approximation to the root, and (if X_0 has been chosen properly) X_{i+1} is a better approximation. When the sequence of approximations converges, it does so at a quadratic rate; that is, each value will have twice as many correct digits (of the root) as its predecessor.

In division, the Newton-Raphson procedure may be used to compute the reciprocal, $1/D$, of the divisor by taking $f(X) = D - 1/X$. Multiplication of this reciprocal by the dividend then yields the quotient: $Q = N \times (1/D)$. The recurrence for the computation of the reciprocal is and the required

---

[6]In practice, the parameters of the algorithm are chosen so that true multiplication is not required.

recurrence is:

$$X_{i+1} = X_i - \frac{D - 1/X_i}{1/X_i^2}$$
$$= 2X_i - DX_i^2$$
$$= X_i(2 - DX_i) \qquad (5.1)$$

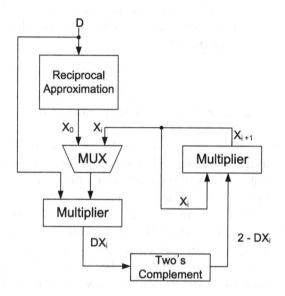

Figure 5.19: Newton-Raphson reciprocator

Where the subtraction from two is a twos complement operation if the operand is assumed to be fractional. A popular version of this scheme concurrently computes approximations to both the quotient and the reciprocal. This is known an *Goldschmidt's algorithm*. Figure 5.19 shows the organization of a Newton-Raphson reciprocator. We have shown two independent multipliers, but, since the multiplications in the above recurrence are independent, it is possible to have the multiplications pipelined through a single multiplier.

Evidently, the number of iterations required to compute $1/D$ to a given precision depends on how good X_0, the starting approximation is, and, in general, its value depends on that of D. If, for example, X_0 is close enough to the root, then only one or two iterations may be required. There are a variety of ways to obtain X_0— table-lookup, piecewise linear interpolation, and so forth—but what they all have in common is that the best approximations are relatively hardware-costly to compute. Nevertheless, as we shall see below, this need not be a problem in residue multiplication, since the "divisor" (the modulus) is usually fixed. There is, however, one problematic aspect of the algorithm: it does not leave a remainder, and computing one is not always easy because the reciprocal value used to obtain the quotient is not exact. The remainder, which is what we are after in modular multiplication, may be computed through an additional multiplication and subtraction: $R = N - Q \times D$.

5.3 Modular multiplication: arbitrary modulus

For the arbitrary moduli, as opposed to the special ones (such as $2^n - 1$ and $2^n + 1$), multipliers based solely on combinational logic are likely to be relatively complex. We shall therefore start with simple multipliers that combine table-lookup with a small amount of combinational logic; such multipliers are most suitable for multiplication relative to a small modulus. Other types of modular multipliers fall into roughly three categories: those that compute $|AB|_m$ by, essentially, computing AB and then reducing that modulo m; those that perform the modular multiplication by a process similar to conventional multiplication but in which each partial product is reduced before being added to other partial products or to a running modular sum; and those that are based on division algorithms, subtractive and multiplicative. We shall discuss examples of all three types.

A key point to keep in mind in what follows is that although the modulus may be arbitrary, it is usually fixed for a given multiplier, since the moduli are usually known at design-time; advantage can be taken of this to make certain simplifications in the designs. We shall assume that the modulus is not a power two; for powers of two, modular multiplication is just conventional multiplication.

5.3.1 Table lookup

One of the first types of multipliers proposed for residue multiplication is that based on squaring [13], as given by the equation

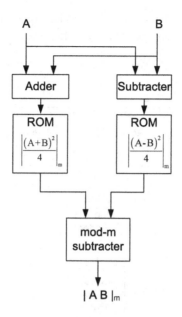

Figure 5.20: Quarter-square modulo-m multiplier

$$A \times B = \frac{(A+B)^2}{4} - \frac{(A-B)^2}{4}$$

As shown in Figure 5.20, modular multiplication on this basis can be implemented directly, using three adders (for the addition and the two subtractions) and two ROMs as lookup tables that produce quarter-squares. Although this type of multiplier is probably more suitable for lookup-table implementation, it may also be implemented as combinational logic: recall that although squaring is essentially a multiplication, it is a specialized one that can be implemented using less logic than would required for a generalized multiplier (Section 5.1). At the other extreme, all three adders may be replaced with lookup tables, to give an all-ROM implementation. Thus, depending on the tradeoffs, there are several possibilities for the implementation of such a modulo-m multiplier. It should nevertheless be noted such "square-law" multiplication cannot be used with all possible values of moduli. The two ROMS in the design above store the values $\left|(A+B)^2|4^{-1}|_m\right|_m$ and $\left|(A-B)^2|4^{-1}|_m\right|_m$, where $|4^{-1}|_m$ is the multiplicative inverse of $|4|_m$; that is, $4|4^{-1}|_m \equiv 1$. But the results in Chapter 2 show that multiplicative inverses exist only if modulus and divisor have no factor in common. Therefore, m here may not have two as a factor.

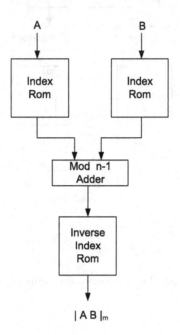

Figure 5.21: Index-calculus modulo-m multiplier

Another design that is suitable for table-lookup implementation can be obtained by using tables of "logarithms" and "anti-logarithms", as one might do in conventional multiplication. This relies on the result of Chapter 1, to the effect that if the modulus m is prime, then every non-zero element of the corresponding set of residues can be generated as a power of some generator, g. Thus if, $\left|A = g^i\right|_m$ and $\left|B = g^j\right|_m$, then their modular product is

$$|AB|_m = \left|g^{|i+j|_{m-1}}\right|_m$$

A design for the corresponding multiplier is shown in Figure 5.21. Three ROMs are used here: one each to obtain i and j from A and B, and one to obtain the final result, $|AB|_m$, from $i + j$. Only one adder is used, for the sum of i and j, and this too could be realized as a ROM. So, depending on the exact-tradeoffs, this may be a more cost-effective design than the preceding one, which nominally requires three adders. An additional point worth noting is that the ROMs here will be smaller than for the basic quarter-square multiplier, since each can be addressed with one bit less. The size of the ROMs can be reduced by decomposing m into several prime submoduli, m_i, and employing a modular adder for each submoduli; this can also increase performance, since the modular adders will then be smaller, and all operate concurrently [20, 21].

This last modular multiplier has two obvious drawbacks. One is that m must be prime. Of course, this is not a problem for commonly used moduli sets—such as $\{2^n - 1, 2^n, 2^n + 1\}$, of which $2^n - 1$ and $2^n + 1$ are prime, and modulo-2^n multiplication is just conventional multiplication—but, on the other hand, for such moduli it is, in general, possible to design better multipliers. The other drawback, a relatively minor one, is that additional logic is required to detect if an operand is zero, so that a result of zero result is produced accordingly.

The designs of Figures 5.20 and 5.21 each requires an addition followed by address-decoding (for the table-lookup). Although, we have implied that these are realized as separate operations, this need not be so. This is one instance where a *sum-addressed memory* can be usefully employed. In contrast with a conventional memory, which takes a single address-input, a sum-addressed memory takes two inputs that it adds and then decodes for use as an address [15]. A performance benefit exists because the addition of the two address-operands does not involve any carry-propagation.

For low-precision moduli, the two multiplier designs just described are probably quite reasonable. But given that the sizes of the required ROMs increase exponentially, relative to an increase in the precisions of the operands, such multipliers may not be suitable for large moduli. For large moduli, combinational-logic multipliers, or multipliers that use largely combinational logic (with, perhaps, a small amount of ROM), are probably better. We discuss these next.

5.3.2 Modular reduction of partial products

A second class of modular multipliers consists of those in which each partial product is reduced modulo m before being added to any other. The basis of such implementations is the equality (Equation 2.1)

$$|X+Y|_m = ||X|_m + |Y|_m|_m$$

This equality may be implemented directly or indirectly. In the former case, each level of the multiplier is a modulo-m carry-save adder [9]. Such an adder is necessarily more complex than a normal carry-save adder, as it has to include a correction term of 0, or $-m$, or $-2m$, according to the result of the initial carry-save addition of two partial products. For example, in the design proposed in [9], the modulo-m carry-save adder consists of five levels of conventional carry-save adders, with intermediary multiplexers to select the correction terms.

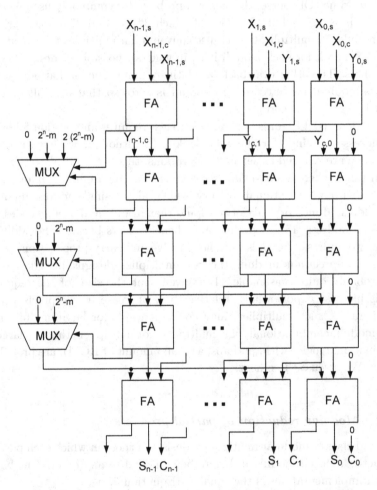

Figure 5.22: Modulo-m carry-save adder

The design is shown in Figure 5.22, in which $*_{s/c,i}$ denotes bit i of the partial-sum or partial-carry corresponding to operand $*$. It is evidently at least five times as costly and eight times as slow as a corresponding level in a conventional multiplier. Nevertheless, it is worth noting that the timing factor can be slightly improved upon: the designers basically implement a 4:2 compressor as a pair of 3:2 compressors in series, but a 4:2 compressor can be implemented in such a way that both the cost and operational delay are less than those for two 3:2 compressors in series [10]. Using the modulo-m carry-save adders, the partial products are summed up in a binary tree whose outputs are a partial-sum and a partial-carry that are then assimilated in a carry-propagate adder. Note that the tree here is much deeper than a Wallace tree in a conventional multiplier. In the proposed design, Booth recoding is not used, although it can be, and so for n-bit operands the partial-product generator simply consists of n^2 AND gates that produce the n^2 bits of the partial products.

```
   5  =     0 0 1 0 1          (A)
  11  =     0 1 0 1 1
            0 0 1 0 1          2⁰ × A × 1
            0 0 1 0 1          2¹ × A × 1
            0 0 0 0 0          2² × A × 0
            0 0 1 0 1          2³ × A × 1
            0 0 0 0 0          2⁴ × A × 0
  55  =   0 0 0 1 1 0 1 1 1
```

Figure 5.23: Paper-and-pencil binary multiplication

In the design just described, each partial product is taken individually and reduced modulo m before being added to another partial product; that is, there are n operands to be added. A different way to proceed is as follows. Consider the paper-and-pencil example of binary multiplication shown in Figure 5.23. If the bits to be added are taken column-wise, then there are $2n$ operands to be added (corresponding to the $2n$ bits of the final result), and these are of precisions $1, 2, 3, \ldots n-1, n, n-1, \ldots, 3, 2, 1$ bits. Each of these $2n$ operands may be reduced modulo m and all reduced column-sums then added up in a tree of carry-save adders and one carry-propagate adder. A design that does exactly this is described in [14] and shown in Figure 5.24.

The first stage of the multiplier, the partial-product generator, consists of just an array of AND gates, but here too Booth recoding is readily applicable. The next stage of the modular multiplier consists of encoders, each of which takes the bits corresponding to a column and produces a *count* of the number of 1s in the column, since it is only these that constitute the column-sums. (Note the simplification in design relative to that of a similar a conventional multiplier.) The third stage of the multiplier consists of a series of ROMs, each of which takes a column-sum and reduces it modulo m. An addition tree then completes the multiplication process; the tree suggested by the designers is that in [11].

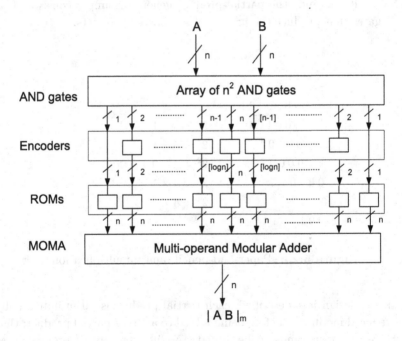

Figure 5.24: Modulo-m multiplier with combinational logic and ROM

One point worth noting about this last multiplier is that the direct column-wise addition is only one of a variety of ways to add up partial products. If one simply starts with the viewpoint that there are only so many partial product bit, and that these may be added in any order, as long as the relative significance of the bits is maintained, then several other arrangements are possible. For conventional multipliers, this yields different CSA-tree organizations [5, 6, 12], any one of which may be substituted for that of Figure 5.24.

5.3.3 Product partitioning

Probably the most direct way to perform the modular multiplication $|AB|_m$ consists, in principle, of a normal unsigned multiplication of A and B followed by a modulo-m reduction of that product. For high performance, the multiplier may be implemented as a tree of carry-save adders (e.g. a Wallace tree or some other structure) and one carry-propagate adder. Appropriate performance-enhancing techniques, such as Booth recoding, may also be used. The high-level design of such a modular multiplier is shown in Figure 5.25. Evidently, some savings in time and cost are possible if the assimilation carry-propagate adder is incorporated into the final modular-reduction logic.

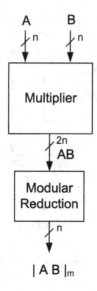

Figure 5.25: Simple combinational modulo-m multiplier

The key issue in such a design is how to reduce the base-product, AB, to the modulo-m product, $|AB|_m$. This can be done by implementing, directly or indirectly, the result in Chapter 2 which states that to compute $|X|_m$, given X, we may partition X into two or more (weighted) pieces and add these in one or more modulo-m adders.

Suppose A, B and m are each represented in n bits and n is as small as possible; that is, $n = 1 + \lfloor \log_2 m \rfloor$. Then AB is $2n$ bits wide, and $m = 2^n - c$, where $1 \leq c < 2^n - 1$. If we split AB into an upper half, U, and a lower half, L, then

$$\begin{aligned}|AB|_m &= |2^n U + L|_m \\ &= \left||2^n U|_m + |L|_m\right|_m \\ &= |cU + L|_m \qquad \text{since } |2^n|_m = c \text{ and } L < m\end{aligned}$$

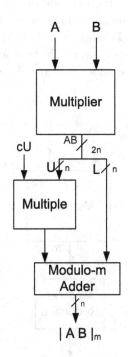

Figure 5.26: Partitioned-operand modulo-m multiplier

Application of this last equation requires another multiplication, but, given that c will generally be small, this operation need not be costly.[7] Thus, corresponding to Figure 5.25, we now have Figure 5.26.

The partitioning above may be extended to even more pieces. A four-piece design is given in [4]. There, AB is partitioned into fours parts: W, of $k+1$ bits; Z, of $n-(k+1)$ bits; Y, of 1 bit; and X of $n-1$ bits, such that

$$P \triangleq AB = 2^{2n-(k+1)}W + 2^n Z + 2^{n-1}Y + X$$

where k is the minimal number of bits required to represent c, and $c = 2^n - m$; that is, $k = 1 + \lfloor \log_2 c \rfloor$. So

$$|AB|_m = \left| \left| 2^{2n-(k+1)}W + 2^n Z + 2^{n-1}Y + X \right| \right|_m$$

$$= \left| \left| 2^{2n-(k+1)}W + 2^{n-1}Y \right|_m + |2^n Z|_m + |X|_m \right|_m$$

Since $2^n = m+c$, we have $|2^n|_m = c$ and $|2^n Z|_m = cZ$. And, since $X < m$, because X is represented in $n-1$ bits and m is represented in n bits, $|X|_m = X$. So

$$|AB|_m = \left| \left| 2^{2n-(k+1)}W + 2^{n-1}Y \right|_m + cZ + X \right|_m$$

EXAMPLE. Suppose $A = 2005 = 01111101010 1_2, B = 3212 = 110010001100_2$, and $m = 4091$. Then $n = 12, c = 5$, and $k = 3$. The product $AB = 01100010010001000111 1100$, and

$$X = 100011111 00_2 = 1148$$
$$Y = 0$$
$$Z = 36$$
$$cZ = 180$$
$$X + cZ = 1328$$
$$\left| 2^{2n-(k+1)}W + 2^{n-1}Y \right|_m = 3589$$
$$|AB|_m = |1328 + 3589|_{4091}$$
$$= 826$$

---

[7]The choice of c is obviously important. For example, if c is a power of two, then the multiplication is just a left-shift. The best cases are $c = -1$ and $c = 1$, which yield the usual special moduli and reduce the entire operation to a single modular subtraction or addition.

END EXAMPLE.

The computation of $\left|\left|2^{2n-(k+1)}W + 2^{n-1}Y\right|_m + cZ + X\right|_m$ is carried out as follows. Let e be the value of $\left|2^{2n-(k+1)}W + 2^{n-1}Y\right|_m$, and suppose that this value is readily available. (The value of k will be small, so the values of e may easily be computed by a small combinational circuit or pre-computed and held in a ROM.)

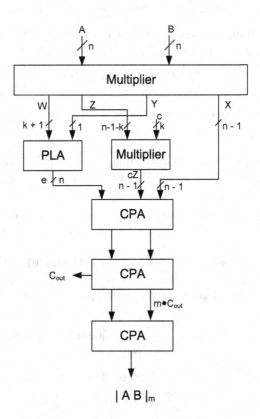

Figure 5.27: Modulo-m product-partitioning multiplier with ROM

The value of e is then added to that of $cZ + X$. Now, $cZ + X < m$, since X is represented in $n - 1$ bits and therefore has a maximum possible value of $2^{n-1} - 1$, Z is represented in $n - (k+1)$ bits and so has a maximum possible value of $2^{n-(k+1)} - 1$, and c is represented in k bits and so has a maximum possible value of $2^k - 1$. Since $e < m$ (by definition), $e + cZ + X < 2m$.

Therefore, after the addition of e to $cZ + X$, at most a single subtraction, of m, will be required to correct the result. In the proposed implementation, it is not convenient to first detect when the modulus has been exceeded and then subtract. Instead, it works slightly better to initially subtract m and then add it back if the resulting value of $e + cZ + X - m$ is negative. Thus the aforementioned ROM/combinational-circuit produces the values of $e - m$ instead of those of e.

A complete design of the modular multiplier just described is shown in Figure 5.27. The values of $e - m$ are negative and are stored in two's complement form. The first multiplier computes the product AB, and the second one computes cZ; both may use any of the high-speed multiplication techniques described in Section 5.1. A carry-save adder is used to reduce the three operands $e - m, cZ$, and X to two; and a carry-propagate adder then assimilates the resulting partial-sum and partial-carry. If the result of this assimilation is negative—and this is indicated by the absence of a carry out of the adder—them m is added back in another carry-propagate adder. Both carry-propagate adders may be of any of the high-performance designs described in Chapter 4. The designs can be modified for higher performance, by having both possible results computed concurrently and one then selected through a multiplexer.

5.3.4 Multiplication by reciprocal of modulus

In principle, modular multiplication can readily be performed by multiplying the two operands, dividing the product by the modulus, and taking as the result of the modular multiplication the remainder from that division. For conventional division there are several algorithms in which the operation is performed by multiplying the dividend by the reciprocal of the divisor. The most well-known of these is based on the Newton-Raphson procedure for finding the roots of non-linear equations (see Section 5.2.2). In the application of this procedure to division, the non-linear function is $f(x) = D - 1/x$, where D is the divisor. The process is generally iterative and requires a starting approximation to the root. In other common variants, the division N/D consists of repeatedly multiplying N by R_i, where R_i are progressively accurate approximations of $1/D$. In either case, a starting approximation, R_0, to the root is required. For conventional division, the divisor is variable and, therefore, so is its approximate reciprocal, which is usually obtained through a combination of table look-up and combinational logic. Furthermore, the number of iterations ultimately required depends on the accuracy of R_0—if it is sufficiently accurate, then a single multiplication is all that is needed—but high accuracy entails high cost. In the case at hand, however, this is not problematic because the divisor, which is the modulus, m, is constant and may therefore be held in a single register or, if it is fixed, even hardwired into the rest of the logic. We next describe modular multiplication based on this approach [3]. $|AB|_m$ is computed by first computing the quotient $Q = \lfloor AB/m \rfloor$, forming the product Qm, and then subtracting that from AB:

$$|AB|_m = AB - mQ$$

Proceeding directly, it might appear that Q can be computed by first computing the reciprocal $1/m$, multiplying that by AB, and then truncating the result. This would be straightforward, but one difficulty is immediately apparent: in general, $1/m$ cannot be represented exactly (because of the finite precision of the computer hardware), which implies that there will be some error in the (supposed) computed value of $1/m$ and therefore in AB/m as well. Thus, instead of Q being computed, the end result will instead be an approximation, Q_a. Nevertheless, if Q_a is computed with enough precision, then Q_a and Q need not differ by more than unity, which leads to two cases: if $Q_a = Q$, then the computed result, $AB - mQ_a$ is $|AB|_m$; otherwise, a correction, by an additional subtraction of m, is necessary. We next show that if A and B are each represented in n bits and $1/m$ is approximated with a precision of at least $2n$ bits, then either $Q_a = Q$ or $Q_a = Q - 1$.

Let r be a k-bit computed approximation to $1/m$. Then for the bounds for the absolute error in r are [8]

$$0 \leq \frac{1}{m} - r < 2^{-k}$$

and the error in the division (i.e. multiplication by AB) has the bounds

$$0 \leq \frac{AB}{m} - rAB < 2^{-k}AB$$

Since A and B are each represented in n bits, $A < 2^n, B < 2^n$, and $AB < 2^{2n}$. Therefore, $2^{-k}AB < 2^{2n-k}$. If the upper bound in the error, ε, is to

---
[8] The upper bound is the maximum absolute error in the last bit on the representation of r: in the worst case, a representation $0 \cdot r_1 r_2 \cdots r_k 111 \cdots 1$ is truncated to $0 \cdot r_1 r_2 \cdots r_k$.

be constrained so as not to exceed unity, then we must have $2^{2n-k} \leq 1$, which implies that $k \geq 2n$. We shall therefore take $k = 2n$.

We now show how to compute the desired value, $Q = \lfloor AB/m \rfloor$. Let Q_a denote the approximation $\lfloor rAB \rfloor$, ε denote the error value $2^{-k}AB$, and $F(x)$ denote the fractional part of x. From the above, we have

$$rAB \leq \frac{AB}{m} < rAB + 2^{-k}AB$$

So

$$\begin{aligned}
Q &= \left\lfloor \frac{AB}{m} \right\rfloor \\
&= \lfloor Q_a + F(rAB) + \varepsilon \rfloor \\
&= Q_a + \lfloor F(rAB) + \varepsilon \rfloor \\
&\stackrel{\triangle}{=} Q_a + \varepsilon_a
\end{aligned}$$

Given that $\varepsilon \leq 1$ and $F(rAB) < 1$, it follows that $0 \leq \varepsilon_a < 2$; that is, $\varepsilon_a = 0$ or $\varepsilon_a = 1$. We may therefore compute $|AB|_m$ as follows.

$$\begin{aligned}
|AB|_m &= AB - mQ \\
&= AB - m\left\lfloor \frac{AB}{m} \right\rfloor \\
&= AB - m(Q_a + \varepsilon_a) \\
&= AB - mQ_a + \varepsilon_a m
\end{aligned}$$

The value of ε_a depends on whether $Q = Q_a$ or $Q = Q_a + 1$, and this can be determined from the value of $AB - mQ_a$. If the latter value is negative, then $Q = Q_a$ and $\varepsilon_a = 0$; otherwise $\varepsilon_a = 1$. In the latter case, m must be subtracted from the computed value, $AB - mQ_a$, to get the correct result. In the implementation, this may be effected by concurrently computing $AB - mQ_a$ and $(AB - mQ_a) - m$ and then selecting one of the two, according to the sign of the latter.

EXAMPLE. Suppose $m = 13, A = 11 = 1011_2$, and $B = 12 = 1100_2$. Then $AB = 10000100_2$, and $k = 8$. $1/B$ truncated to eight fractional bits is $0 \cdot 00010011$. From which we get

$$\begin{aligned}
Q_a &= 01001 \\
\varepsilon_a &= 1 \\
|AB|_m &= 10000100 - (1001 \times 1101) - (1 \times 1101) \\
&= 0010 \\
&= 2
\end{aligned}$$

So $|11 \times 12|_{13} = 2$.

END EXAMPLE

The design of the modular multiplier is as shown in Figure 5.28. The first multiplier computes AB. The second multiplier computes Q_a, the integer approximation to the quotient, as the product of AB and (the approximation to) $1/m$. (This is the nominal "Newton-Raphson" phase.) And the third multiplier computes mQ_a. All that remains then is the subtraction (which is also an addition) to get the reminder of the division; that is, the computation of $AB - mQ_a$. Since the computation of Q_a is not exact, and $AB - mQ_a$ can be too large by m, the two adders compute $AB - mQ_a$ and $(AB - mQ_a) - m$, one of which is then selected as the result, according to the sign of the latter. In design of [3], all three multipliers are full multipliers, which means that if they are high-speed ones, then each consists of rows of carry-save adders and a carry-propagate adder, with the two adders being full carry-propagate adders. Therefore, most of the operational delay will be due to the carry-propagate adders. To speed up the process, all of the carry-propagate adders in the multipliers can be eliminated, the intermediate result left in carry-save form, and final assimilation of partial-carry and partial-sum done in the last two adders. The extra cost is an increase in the numbers of some of the interconnections and another level of carry-save adders in each multiplier. We leave it to the diligent reader to work out the details.

5.3.5 Subtractive division

Another straightforward design for a hardware unit to compute $|AB|_m$ from A, B and m, would be one that consists of the use one of the multipliers described in Section 5.1 to compute AB, use one of the dividers of Section 5.2 to divide AB by m, and then take the remainder from that division. But there is an evident similarity between the multiplier of Figure 5.3 and the divider of Figure 5.16, as well as between the multiplier of Figure 5.10 and the divider of Figure 5.18. These similarities suggest that the multiplication and the division can be carried out in a single unit.

In multiplication a partial product may be reduced, relative to the modulus, before being added to another (Equation 2.1). In the corresponding step in subtractive division, a multiple of the divisor (which in this case is the modulus) is subtracted. We may therefore have a combined unit in which each cycle consists of the formation of a multiplicand-multiple followed by the subtraction of a divisor-multiple. Thus each carry-save adder in the corresponding multiplier or divider gets replaced With two carry-save adders—one for the addition and one for the subtraction.

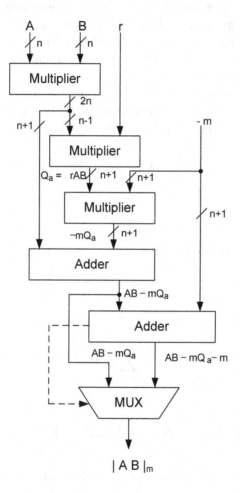

Figure 5.28: Modulo-m reciprocal multiplier

5.4 Modular multiplication: modulus $2^n - 1$

Multiplication modulo-$(2^n - 1)$ can be carried out easily by reducing each partial product as it is generated and added. Since the additions of partial products will normally be done in carry-save adders, and the final partial-carry and partial-sum from these adders have to be assimilated in a carry-propagate adder, the final reduction may be realized by making the assimilating adder a modulo-$(2^n - 1)$ carry-propagate adder. We next describe an algorithm for modulo-$(2^n - 1)$ multiplication and then discuss various possible implementations. We shall initially assume straightforward multiplication; that is, that no speed-up techniques, such as Booth's Algorithm, are used.

The modular product $|AB|_{2^n-1}$ nominally consists of modulo-$(2^n - 1)$ sum of n partial products, each of which is AB_i shifted left by i bits (i.e. multiplied by 2^i), where B_i is the ith bit of B:

$$|AB|_{2^n-1} = \left| \sum_{i=0}^{n-1} 2^i AB_i \right|_{2^n-1}$$

$$\triangleq \sum_{i=0}^{n-1} P^{(i)}$$

And the ith partial product, $P^{(i)}$, is

$$P^{(i)} = \left| 2^i \sum_{j=0}^{n-1} P_j^{(i)} 2^j \right|_{2^n-1}$$

$$= \left| 2^i \sum_{j=0}^{n-1} A_j B_i 2^j \right|_{2^n-1}$$

$$= \left| 2^i B_i (A_{n-1} 2^{n-1} + A_{n-2} 2^{n-2} + \cdots A_0 2^0) \right|_{2^n-1}$$

$$= \left| 2^{i-1} A_{n-1} B_i 2^n + 2^{i-2} A_{n-2} B_i 2^n + \cdots + A_{n-i} B_i 2^n + 2^{n-1} A_{n-i-1} B_i \right.$$
$$\left. + 2^{n-2} A_{n-i-2} B_i + \cdots + 2^i A_0 B_i \right|_{2^n-1}$$

$$= \left| \left| 2^{i-1} A_{n-1} B_i 2^n \right|_{2^n-1} + \left| 2^{i-2} A_{n-2} B_i 2^n \right|_{2^n-1} + \cdots + \left| A_{n-i} B_i 2^n \right|_{2^n-1} \right.$$
$$\left. + \left| 2^{n-1} A_{n-i-1} B_i \right|_{2^n-1} + \left| 2^{n-2} A_{n-i-2} B_i \right|_{2^n-1} + \cdots \right.$$
$$\left. + \left| 2^i A_0 B_i \right|_{2^n-1} \right|_{2^n-1}$$

Now $|2^n|_{2^n-1} = 1$. Also, for each term, $2^k A_j$, above, we have $|2^k A_j B_i|_{2^n-1} = 2^k A_j B_i$, since $A_j B_i$ is 0 or 1 and $2^k < 2^n - 1$. So

$$P^{(i)} = \left|2^{i-1}A_{n-1}B_i + 2^{i-2}A_{n-2}B_i + \cdots + A_{n-i}B_i + \right.$$
$$\left. 2^{n-1}A_{n-i-1}B_i + 2^{n-2}A_{n-i-2}B_i + \cdots + 2^i A_0 B_i\right|_{2^n-1}$$

$$= \left|2^{n-1}A_{n-i-1}B_i + 2^{n-2}A_{n-i-2}B_i + \cdots + 2^i A_0 B_i + 2^{i-1}A_{n-1}B_i + \right.$$
$$\left. 2^{i-2}A_{n-2}B_i + \cdots + A_{n-i}B_i\right|_{2^n-1}$$

$$= \left|(2^{n-1}A_{n-i-1} + 2^{n-2}A_{n-i-2} + \cdots + 2^i A_0 + 2^{i-1}A_{n-1} + 2^{i-2}A_{n-2} \right.$$
$$\left. + \cdots + A_{n-i})B_i\right|_{2^n-1}$$

The expression by which B_i is multiplied is the numerical value represented by the binary pattern $A_{n-i-1}A_{n-i-2}\cdots A_0 A_{n-1}A_{n-2}\cdots A_{n-i}$, which is just the result of an i-bit cyclic shift of the representation of A. Since $A < 2^n - 1$, there must be some k for which $A_k = 0$. Therefore, the value of the expression must also be less than $2^n - 1$ and so

$$P^{(i)} = 2^{n-1}A_{n-i-1}B_i + 2^{n-2}A_{n-i-2}B_i + \cdots + 2^i A_0 B_i 2^{i-1}A_{n-1}B_i +$$
$$2^{i-1}A_{n-2}B_i + \cdots + A_{n-i}B_i$$

(One may here view the formation of the partial products as a sequence of cyclic convolutions [2], but, it is not a view that is particularly enlightening.) Figure 5.29 shows an example of a modulo-$(2^n - 1)$ multiplication. Note that adding the end-around-carries do not involve any extra additions: each such carry is simply the carry-in to the adder at that level, as shown in Figure 5.30. The most noteworthy aspect of this example is compaction of the partial-product array—it is now of size $n \times n$, in contrast with the $n \times 2n$ of a conventional array. A corresponding multiplier-design is shown in Figure 5.30.

The method just described for binary multiplication modulo-$(2^n - 1)$ is easily extensible to higher radices, either by the use of Booth recoding (Table 5.3) or by the use of high-radix digits. In the latter case, it is convenient to select a radix that is a power of two; then the conversion between binary and larger radices is simply a matter of grouping bits of the operands.

14 x 13 (mod 15) = 2

```
   1110    (14)
   1101    (13)
   1110    P⁽⁰⁾
   0000    P⁽¹⁾
   1110
   1011    P⁽²⁾
   1001
      1    End-Around-Carry
   1010
   0111    P⁽³⁾
   0001
      1    End-Around-Carry
   0010    (2)
```

Figure 5.29: Example multiplication modulo $2^n - 1$ $(n = 5)$

A high-radix modulo-$(2^n - 1)$ multiplier of the latter type has been proposed in [2]. We briefly describe the basic principles involved in that design. Suppose the chosen radix is 2^k, where $n = km$ for some m. Then the n-bit operands A and B are each partitioned into m, k-bit, blocks, $\mathbf{A}_{m-1}, \mathbf{A}_{m-2}, \ldots, \mathbf{A}_0$ and $\mathbf{B}_{m-1}, \mathbf{B}_{m-2}, \ldots, \mathbf{B}_0$, such that

$$A = \mathbf{A}_0 + \mathbf{A}_1 2^k + \mathbf{A}_2^{2k} + \cdots + \mathbf{a}_{m-1} 2^{(m-1)k}$$
$$B = \mathbf{B}_0 + \mathbf{B}_1 2^k + \mathbf{B}_2^{2k} + \cdots + \mathbf{b}_{m-1} 2^{(m-1)k}$$

Then

$$|AB|_{2^n-1} = \left| \sum_{i=0}^{m-1} \left(2^{ik} \mathbf{B}_i \sum_{j=0}^{m-1} 2^{jk} \mathbf{A}_j \right) \right|_{2^n-1}$$
$$\stackrel{\Delta}{=} \sum_{i=0}^{m-1} P^{(i)}$$

and a similar partial-product reduction process to that above for the radix-2 algorithm yields similar results here for radix 2^k:

Multiplication

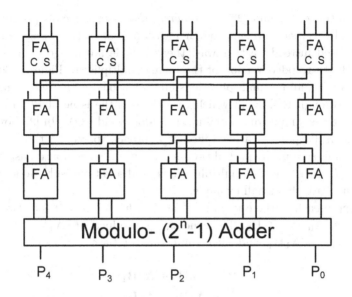

Figure 5.30: Parallel modulo-$(2^n - 1)$ multiplier

$$P^{(i)} = \left| 2^{ik} \mathbf{B}_i \left(\mathbf{A}_0 + \mathbf{A}_1 2^k + \mathbf{a}_2 2^{2k} + \cdots + \mathbf{A}_{m-i-1} 2^{(m-i-1)k} + \mathbf{A}_{m-i} 2^{(m-i)k} \right. \right.$$
$$\left. \left. + \mathbf{A}_{m-i+1} 2^k + \cdots + \mathbf{A}_{m-2} 2^{(m-2)k} + \mathbf{A}_{m-1} 2^{(m-1)k} \right) \right|_{2^n - 1}$$

$$= \left| \mathbf{B}_i \left(\mathbf{A}_0 2^{ik} + \mathbf{A}_1 2^{(i+1)k} + \mathbf{a}_2 2^{(i+2)k} + \cdots + \mathbf{A}_{m-i-1} 2^{(m-1)k} + \mathbf{A}_{m-i} 2^{mk} \right. \right.$$
$$\left. \left. + \mathbf{A}_{m-i+1} 2^{(m+1)k} + \cdots + \mathbf{A}_{m-2} 2^{(m+i-2)k} + \mathbf{A}_{m-1} 2^{(m+i-1)k} \right) \right|_{2^n - 1}$$

$$= \left| \mathbf{B}_i \left(\mathbf{A}_0 2^{ik} + \mathbf{A}_1 2^{(i+1)k} + \mathbf{A}_2 2^{(i+2)k} + \cdots + \mathbf{A}_{m-i-1} 2^{(m-1)k} \right. \right.$$
$$\left. \left. + \mathbf{A}_{m-i} + \mathbf{A}_{m-i+1} 2^k + \cdots + \mathbf{A}_{m-2} 2^{(i-2)k} + \mathbf{A}_{m-1} 2^{(i-1)k} \right) \right|_{2^n - 1}$$

$$= \left| \mathbf{B}_i \left(\mathbf{A}_{m-i-1} 2^{(m-1)k} + \cdots + \mathbf{a}_1 2^{(i+1)k} + \mathbf{A}_0 2^{ik} + \mathbf{A}_{m-1} 2^{(i-1)k} \right. \right.$$
$$\left. \left. + \mathbf{A}_{m-2} 2^{(i-2)k} + \cdots + A_{m-i} \right) \right|_{2^n - 1}$$

Once again, we see that the ith partial product is formed by multiplying the result of an i-digit (i.e a ki-bit) cyclic shift of the multiplicand—from $\mathbf{A}_{m-1} \mathbf{A}_{m-2} \cdots \mathbf{A}_{m-i} \cdots \mathbf{A}_0$ to $\mathbf{A}_{m-i-1} \cdots \mathbf{A}_1 \mathbf{A}_0 \; \mathbf{A}_{m-1} \mathbf{a}_{m-2} \cdots \mathbf{A}_{m-i}$— with a digit of the multiplier.

Assuming a parallel multiplication structure, once the partial products have been formed, a direct way to obtain the final product is through additions in a tree of modulo-$(2^n - 1)$ adders. Of course, a tree of appropriately connected carry-save adders, together with a final carry-propagate adder, may be used instead. Either way, there is little advantage to be gained over a Booth-recoded version of the binary multiplier of Figure 5.30. the other extreme, all the multiplications and additions of Figure 5.32 may be realized through ROMS, in which case there is the issue of ROM size to consider, or as a mixture of combinational-logic and ROM. In [2], however, the authors propose a rather different approach, based on the fact that each column-sum in the partial-product array can be computed using only squaring (rather than full multiplications) and additions/subtractions. We show this through a small example.

Suppose each of the operands, A and B, has been split into two k-bit digits— $\mathbf{A_1 A_0}$ and $\mathbf{B_1 B_0}$. Then the partial products are $\mathbf{A_1 B_0} 2^k + \mathbf{A_0 B_0}$ and $\mathbf{A_0 B_1} 2^k + \mathbf{A_1 B_1}$, and the column sums, P_i, are

$$P_0 = \mathbf{A_0 B_0} + \mathbf{A_1 B_1}$$
$$P_1 = \mathbf{A_1 B_0} + \mathbf{A_0 B_1}$$

(Note that the column sums must be added, with appropriate significance-shifting, and then reduced modulo $2^n - 1$ to get the final result; that is, the final result is $|P_1 2^k + P_0|_{2^n - 1}$.) Now define

$$a = \mathbf{A_0} + \mathbf{A_1} + \mathbf{b_0} + \mathbf{B_1}$$
$$b = \mathbf{A_0} + \mathbf{A_1} - \mathbf{b_0} - \mathbf{B_1}$$
$$c = \mathbf{A_0} - \mathbf{A_1} + \mathbf{b_0} - \mathbf{B_1}$$
$$d = \mathbf{A_0} - \mathbf{A_1} - \mathbf{b_0} + \mathbf{B_1}$$

Then

$$P_0 = \left(a^2 - b^2 + c^2 - d^2\right)/8$$
$$P_1 = a^2 - b^2 - c^2 + d^2/8$$

(where the division by 8 is not a real division). The corresponding hardware structure is shown in Figure 5.31. In [2], it is proposed that ROMs be used for the squaring operations, but there is no indication of how the final result is to be computed from the P_is. The best structure for the later is evidently a carry-save tree, as we have assumed in Figure 5.30.

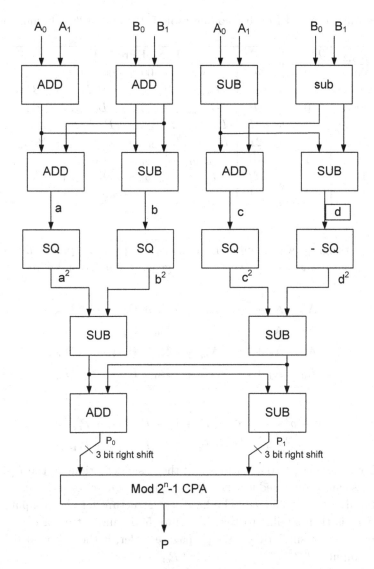

Figure 5.31: High-radix modulo-$2^n - 1$ multiplier

Table 5.3: Radix-4 Booth recoding table for modulo-$(2^n - 1)$ multiplier

$B_{2i+1,2i}$	B_{2i-1}		Partial Product
00	0	$+0$	$00\cdots000\cdots0$
00	1	$+B$	$B_{n-2i-1}B_{n-2i-1}\cdots B_0 B_{n-1}\cdots B_{n-2i}$
01	0	$+B$	$B_{n-2i-1}B_{n-2i-1}\cdots B_0 B_{n-1}\cdots B_{n-2i}$
01	1	$+2B$	$B_{n-2i-2}B_{n-2i-1}\cdots B_0 B_{n-1}\cdots B_{n-2i-1}$
10	0	$-2B$	$\overline{B}_{n-2i-2}\overline{B}_{n-2i-1}\cdots \overline{B}_0\overline{B}_{n-1}\cdots \overline{B}_{n-2i-1}$
10	1	$-B$	$\overline{B}_{n-2i-1}\overline{B}_{n-2i-1}\cdots \overline{B}_0\overline{B}_{n-1}\cdots \overline{B}_{n-2i}$
11	1	$-B$	$\overline{B}_{n-2i-1}\overline{B}_{n-2i-1}\cdots \overline{B}_0\overline{B}_{n-1}\cdots \overline{B}_{n-2i}$
11	1	-0	$00\cdots000\cdots0$

It is straightforward, but quite tedious, to apply the procedure above to partitions of more than two digits. In general, if a, b, c and d are defined as

$$a = \mathbf{A}_0 + \mathbf{A}_1 + \cdots + \mathbf{A}_{m-1} + \mathbf{B}_0 + \mathbf{B}_1 + \cdots + \mathbf{B}_{m-1}$$
$$b = \mathbf{A}_0 + \mathbf{A}_1 + \cdots + \mathbf{A}_{m-1} - \mathbf{B}_0 + \mathbf{B}_1 - \cdots - \mathbf{B}_{m-1}$$
$$c = \mathbf{A}_0 - \mathbf{A}_1 + \cdots - \mathbf{A}_{m-1} + \mathbf{B}_0 - \mathbf{B}_1 + \cdots - \mathbf{B}_{m-1}$$
$$c = \mathbf{A}_0 - \mathbf{A}_1 + \cdots - \mathbf{A}_{m-1} - \mathbf{B}_0 + \mathbf{B}_1 - \cdots + \mathbf{B}_{m-1}$$

then

$$a^2 - b^2 + c^2 - d^2 = P_0 + P_2 + P_4 + \cdots + P_{m-2}$$
$$a^2 - b^2 - c^2 + d^2 = P_1 + P_3 + P_5 + \cdots + P_{m-1}$$

and from these last two equations and the basic definitions of the P_is, expressions can be obtained for each P_i as a sum of squares of terms involving just the digits of A and B. The hardware structure for the computation of the P_i is then similar to that of Figure 5.31, and a tree of carry-save adders and an assimilating carry-propagate adder, is then appended to finally compute $\left|2^{(m-1)k}P_{m-1} + 2^{(m-2)k}P_{m-2} + \cdots + P_0\right|_{2^n-1}$.

Nevertheless, as indicated above, the performance of the resulting high-radix structure is unlikely to be better than the more direct radix-2 approach combined with Booth recoding. In particular, it will be noted that all the adders used in the computation of the Ps will be carry-propagate adders. Furthermore, there is little that can be done to improve this: partitioning the operands into fewer digits results in wider and slower adders, and more digits simply increases the depth of the computation tree. Some minor improvements are possible though: for example, the adders just before the ROMs can be eliminated by implementing the ROMs as sum-addressable memories [15].

5.5 Modular multiplication: modulus $2^n + 1$

Multiplication modulo $2^n + 1$ is especially important, given its application in cryptography and other areas, but it is considerably more difficult than multiplication modulo $(2^n - 1)$. In what follows we shall discuss one approach that is among the best of the existing ones.

Given two operands, A and B, the modulo-$(2^n + 1)$ product is given by

$$|AB|_{2^n+1} = \left| \sum_{i=0}^{n-1} 2^i B_i A \right|_{2^n+1}$$

$$\stackrel{\triangle}{=} \left| \sum_{i=0}^{n-1} P^{(i)} \right|_{2^n+1}$$

Each partial product, $P^{(i)} \stackrel{\triangle}{=} 2^i A B_i$, is nominally represented in $2n$ bits, and we may therefore split (the representation of)[9] $2^i A$, which is

$$\overbrace{00\cdots0}^{n-i\ 0s} A_{n-1} A_{n-2} \cdots A_0 \overbrace{00\cdots0}^{i\ 0s}$$

into two n-bit pieces U and L:

$$L = A_{n-i-1} A_{n-i-2} \cdots A_0 \overbrace{00\cdots0}^{i\ 0s}$$

$$U = \overbrace{00\cdots0}^{n-i\ 0s} A_{n-1} A_{n-2} \cdots A_i$$

Then $2^i A = 2^n U + L$, and so

$$P^{(i)} = \left|\ |B_i 2^n U|_{2^n+1} + |B_i L|_{2^n+1}\ \right|_{2^n+1}$$
$$= \left|\ |B_i L|_{2^n+1} - |B_i U|_{2^n+1}\ \right|_{2^n+1}$$
$$= |B_i L - B_i U|_{2^n+1}$$

since $|2^n|_{2^n+1} = -1$, L and U are each represented in n bits, and B_i is 0 or 1 (i.e $B_i L < 2^n$ and $B_i U < 2^n$).

---
[9]We remind the reader that for ease of presentation, we do not always make a distinction between a *number* and its *representation*.

Now, from Section 1.1, the numeric value, $\overline{U}$, of the 1s complement of a number U is

$$\overline{U} = 2^n - 1 - U \tag{5.2}$$

so

$$\begin{aligned}
|-U|_{2^n+1} &= \left|-2^n + 1 + \overline{U}\right|_{2^n+1} \\
&= \left|-(2^n + 1) + 2 + \overline{U}\right|_{2^n+1} \\
&= \left|2 + \overline{U}\right|_{2^n+1} \\
&= \overline{U} + 2
\end{aligned}$$

Therefore

$$P^{(i)} = |B_i(L - U)|_{2^n+1}$$
$$|B_i(L + \overline{U} + 2)|_{2^n+1}$$

$$= |B_i(A_{n-i-1}A_{n-i-2}\cdots A_0 \overbrace{00\cdots 0}^{i\ 0s} + \overbrace{11\cdots 1}^{n-i\ 1s} \overline{A}_{n-1}\overline{A}_{n-2}\cdots \overline{A}_{n-i} + 10_2)|_{2^n+1}$$

This expression can then be simplified through further algebraic manipulation involving certain terms, of value zero, introduced for just that purpose [16]:

$$P^{(i)} = |B_i \cdot (A_{n-i-1}A_{n-i-2}\cdots A_0 \overbrace{00\cdots 0}^{i\ 0s} + \overbrace{11\cdots 1}^{n-i\ 1} \overline{A}_{n-1}\overline{A}_{n-2}\cdots \overline{A}_{n-i}$$
$$10_2)\overline{B}_i(-00\cdots 0_2)|_{2^n+1}$$

$$= |B_i(A_{n-i-1}A_{n-i-2}\cdots A_0 00\cdots 0 + 11\cdots 1\overline{A}_{n-1}\overline{A}_{n-2}\cdots \overline{A}_{n-i} + 10_2)$$
$$+ \overline{B}_i(11\cdots 1 + 10_2)|_{2^n+1}$$

$$= |B_i(A_{n-i-1}A_{n-i-2}\cdots A_0 00\cdots 0 + 11\cdots 1\overline{A}_{n-1}\overline{A}_{n-2}\cdots \overline{A}_{n-i}+10_2)$$
$$+\overline{B}_i(00\cdots 011\cdots 1 + 11\cdots 100\cdots 0 + 10_2)|_{2^n+1}$$

$$= |B_i(A_{n-i-1}A_{n-i-2}\cdots A_0 00\cdots 0 + 00\cdots 0\overline{A}_{n-1}\overline{A}_{n-2}\cdots \overline{A}_{n-i})$$
$$+\overline{B}_i(00\cdots 011\cdots 1) + (B_i + \overline{B}_i)(11\cdots 100\cdots 0 + 1)|_{2^n+1}$$

$$= |B_i(A_{n-i-1}A_{n-i-2}\cdots A_0\overline{A}_{n-1}\overline{A}_{n-2}\cdots \overline{A}_{n-i}) +$$
$$\overline{B}_i(00\cdots 011\cdots 1 + 1)(11\cdots 100\cdots 0 + 1)|_{2^n+1}$$

Now the numerical value represented by the binary $\overbrace{11\cdots 1}^{n-i\ 1s}\overbrace{00\cdots 0}^{i\ 0s}+1$ is

$$\sum_{j=i}^{n-1} 2^i + 1 = 2^n - 2^i + 1$$

So

$$\left|\sum_{i=1}^{n-1}(2^i+1)\right|_{2^n+1} = |(2^n+1) - 2^i|_{2^n+1}$$
$$= -2^i$$

and over the n partial products

$$\sum_{i=0}^{n-1}\left(\overbrace{11\cdots 1}^{n-i\ 1s}\overbrace{00\cdots 0}^{i\ 0s}+1\right) = \left|\sum_{i=0}^{n-1} -2^i\right|_{2^n+1}$$
$$= |n(2^n+1) - (2^n+1) + 2|_{2^n+1}$$
$$= 2$$

Therefore it is sufficient to modify the partial products to

$$P^{(i)} = |B_i \cdot (A_{n-i-1}A_{n-i-2}\cdots A_0\overline{A}_{n-1}\overline{A}_{n-2}\cdots \overline{A}_{n-i})$$
$$+\overline{B}_i \cdot (00\cdots 011\cdots 1 + 1)|_{2^n+1}$$

with a 2 added in at the end. The design of the corresponding n-bit modulo-(2^n+1) carry-save adder is obtained by taking a normal carry-save adder and adding the inverse of the carry-out as a carry-in. This follows from the following reasoning. From the last expression for $P^{(i)}$, each carry-save adder needs to add in a 1. Also the two operands, X and Y, that are input to the carry-save adder, and their sum, S, will satisfy the conditions $0 < X < 2^n$, $0 < Y < 2^n$, and $0 < S < 2^{n+1}$. (What we need to compute is $S + 1$.) If $0 < S < 2^n$, then there is no carry-out, and the result is in the right range. Otherwise, i.e. if $S \geq 2^n$, then there is a carry-out. In the latter case, $S = 2^n + c$, where $0 < c < 2^n$; so there is a carry-out and c is represented by the remaining n bits. Here

$$|S+1|_{2^n+1} = |(2^n+1) + c|_{2^n+1}$$
$$= c$$

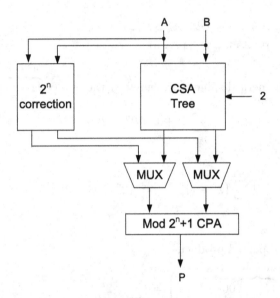

Figure 5.32: Parallel modulo $2^n + 1$ multiplier

So if there is a carry-out, C_{out}, then no further action is required: the result is already $S + 1$. But if there is no carry-out then a 1 must be added to obtain $S + 1$. Because the multiplier array has several levels, the addition may be accomplished by adding C_{out} from one carry-save adder as C_{in} to the carry-save adder at the next level. The 1 from the last level of carry-save addition is added in the final, assimilating modulo-$(2^n + 1)$ carry-propagate adder.

Lastly, the value 2^n, which is represented by $00 \cdots 0$ requires special handling. Suppose A is 2^n but B is not. Then

$$|AB|_{2^n+1} = |2^n B|_{2^n+1}$$
$$= |-B|_{2^n+1}$$
$$= |\overline{B} + 2|_{2^n+1} \quad \text{By Equation 5.1}$$

Similarly, if B is 2^n but A is not, then $|AB|_{2^n+1} = |\overline{A}+2|_{2^n+1}$. And if both are equal to 2^n, then $|AB|_{2^n+1} = 1$. Since the assimilating carry-propagate adder adds in a 1 (C_{in} above), in this case the inputs to the adder that are required to get the correct results (i.e. $\overline{B}+2$, or $\overline{A}+2$, or 1) are $\overline{B}+1$, or $\overline{A}+1$, or 0. In ⟨partial-carry, partial-sum form⟩ these are $(\overline{B}, 1)$, or $(\overline{A}, 1)$, or $(0,0)$. The final design of the multiplier is therefore as shown in Figure 5.32.

Booth recoding can be used in the modulo-$(2^n + 1)$ multiplier, but its application here is untidy in that a correction, T, that depends on the multiplier-operand, must now be added and the constant 2 above replaced with 1.

Table 5.4: Radix-4 Booth recoding table for modulo-$(2^n + 1)$ multiplier

$B_{2i+1,2i}$	B_{2i-1}		Partial Product
00	0	$+0$	$00 \cdots 011 \cdots 1$
00	1	$+B$	$B_{n-2i-1}B_{n-2i-1} \cdots B_0 \overline{B}_{n-1} \cdots \overline{B}_{n-2i}$
01	0	$+B$	$B_{n-2i-1}B_{n-2i-1} \cdots B_0 \overline{B}_{n-1} \cdots \overline{B}_{n-2i}$
01	1	$+2B$	$B_{n-2i-2}B_{n-2i-1} \cdots B_0 \overline{B}_{n-1} \cdots B_{n-2i-1}$
10	0	$-2B$	$\overline{B}_{n-2i-2}\overline{B}_{n-2i-1} \cdots \overline{B}_0 B_{n-1} \cdots B_{n-2i-1}$
10	1	$-B$	$\overline{B}_{n-2i-1}\overline{B}_{n-2i-1} \cdots \overline{B}_0 B_{n-1} \cdots B_{n-2i}$
11	1	$-B$	$\overline{B}_{n-2i-1}\overline{B}_{n-2i-1} \cdots \overline{B}_0 B_{n-1} \cdots B_{n-2i}$
11	1	-0	$11 \cdots 100 \cdots 0$

For 2-bit recoding, the bits of T are

$$T_0 = \overline{B}_1 \overline{B}_0 \overline{B}_{n-1} + B_1 \overline{B}_0 B_{n-1} + B_0 \overline{B}_{n-1}$$
$$T_1 = \overline{B}_1 + \overline{B}_0 \overline{B}_{n-1}$$
$$T_{2i} = \overline{B}_{2i+1} \overline{B}_{2i} + B_{2i+1} \overline{B}_{2i} + \overline{B}_{2i} B_{2i-1}$$
$$T_{2i+1} = \overline{B}_{2i+1}$$

The recoding table is then as shown in Table 5.4.

As indicated in Chapter 4, modulo-$(2^n + 1)$ arithmetic is also sometimes done indirectly, via diminished-one representations; this simplifies hardware designs. We next show this for multiplication.

Let $\widetilde{A}$ denote the diminished-one representation of A; that is, $\widetilde{A} = A - 1$.

Then
$$2\widetilde{X} = \widetilde{X+X}$$
$$= 2\widetilde{X} + 1$$

and adding X to itself $2^i - 1$ times yields
$$\widetilde{2^i X} = 2^i \widetilde{X} + 2^i 1$$
$$= 2^i(\widetilde{X} + 1) - 1$$

Now, the product $\widetilde{AB}$ is just A added to itself $B - 1$ times, so

$$|\widetilde{AB}|_{2^n+1} = \left|\widetilde{A}(\widetilde{B} + 1) + \widetilde{B}\right|_{2^n+1}$$

$$= \left|\sum_{i=0}^{n-1} \widetilde{2^i A}(\widetilde{B}_i + 1) + \sum_{i=0}^{n-1} 2^i \widetilde{B}_i\right|_{2^n+1}$$

$$= \left|\sum_{i=1}^{n-1} \widetilde{B}_i(2^i \widetilde{A} + 2^i - 1) + \widetilde{A}(\widetilde{B}_0 + 1) + \sum_{i=0}^{n-1} 2^i \widetilde{B}_i\right|_{2^n+1}$$

$$= \left|\sum_{i=1}^{n-1} 2^i \widetilde{A} \widetilde{B}_i + \sum_{i=1}^{n-1} 2^i \widetilde{B}_i + \widetilde{A}(\widetilde{B}_0 + 1) + \widetilde{B}_0\right|_{2^n+1}$$

To progress further, we note that since multiplication by 2^i is an i-bit left shift, the representation of $2^i \widetilde{A}$ is

$$\widetilde{A}_{n-1}\widetilde{A}_{n-2}\cdots \widetilde{A}_{n-k}\widetilde{A}_{n-i-1}\widetilde{A}_{n-i-2}\cdots \widetilde{A}_0 \overbrace{00\cdots 0}^{i\ 0s}$$

If we split this into two parts U, consisting of the most significant n bits, and L, consisting of the least significant $n - i$ bits, then

$$\left|2^i \widetilde{A} + 2^i - 1\right|_{2^n+1} = \left|2^n U + L + 2^i - 1\right|_{2^n+1}$$
$$= \left|L + (2^i - 1 - U)\right|_{2^n+1} \quad \text{since } |2^n U|_{2^n+1} = -U$$
$$= L + \overline{U}$$

Therefore, the representation of $2^i \widetilde{A}$ is $\widetilde{A}_{n-i-1}\widetilde{A}_{n-i-2}\cdots \widetilde{A}_0 \overline{\widetilde{A}_{n-1}}\ \overline{\widetilde{A}_{n-2}}\cdots \overline{\widetilde{A}_{n-i}}$, and it is obtained simply by k-bit left rotation with a complementation of the bits rotated out and around.

5.6 Summary

Implementations of modular multiplication are generally of one of two types: those in which the primary operation is viewed as multiplication, and the main task is to perform modular reductions of partial products and their sums, and those that proceed directly from the definition of a residue, and the main task is the modular reduction of a base product. In latter case, the reduction is by the pseudo-division of a product by a modulus, with the remainder taken as the sought value; in such implementations, it is the essence of division algorithms that is employed, and an actual divider is not realized. The many varieties of multipliers and dividers give rise to a large number of possible implementations, and, as the case with all RNS operations, designs are simpler and more efficient when the modulus is not arbitrary.

References

(1) V. Paliouras, K. Karagianni, and T. Stouraitis. 2001. A low complexity combinatorial RNS multiplier. *IEEE Transactions on Circuits and Systems–II*, 48(7):6675–683.

(2) A. Skavantzos and P.B. Rao. 1992. New multipliers modulo $2^N - 1$. *IEEE Transactions on Computers*, 41(8):957–961.

(3) G. Alia and E. Martinelli. 1991. A VLSI modulo m multiplier. *IEEE Transactions on Computers*, 40(7):873–877.

(4) A. Hiasat. 2000. New efficient structure for a modular multiplier for RNS. *IEEE Transactions on Computers*, 49(2):170–173.

(5) A. R. Omondi. 1993. *Computer Arithmetic Systems*. Prentice-Hall, UK.

(6) B. Parhami. 2000. *Computer Arithmetic*. Oxford University Press, UK.

(7) M. D. Ercegovac and T. Lang. 1987. On-the-fly conversion of redundant into conventional representations. *IEEE Transactions on Computers*, 36(7):895–897.

(8) R.K. Kolagotla, W.R. Griesbach, and H.R. Srinivas. 1998. VLSI implementation of 350 MHz 0.35 micron 8-bit merged squarer. *Electronics Letters*, 34(1):47–48.

(9) K. M. Elleithy and M.A. Bayoumi. 1995 A systolic architecture for modulo multiplication. *IEEE Transactions on Circuits and Systems*

– *II*, 42(11):725–729.

(10) N. Ohkubo et al. 1994. A 3.nns 54×54b multiplier using pass transistor multiplexer. *IEEE Journal of Solid-State Circuits*:251-257.

(11) S. J. Piestrak. 1993. Design of residue generators and multi-operand modular adders using carry-save adders. *IEEE Transactions on Computers*, 43(1):68–77.

(12) M. J. Flynn and S. F. Oberman. 2001. *Advanced Computer Arithmetic*. Wiley, New York.

(13) M. A. Soderstrand. 1983. A new hardware implementation of modulo adders for residue number systems. In: *Proceedings, 26th Midwest Symposium on Circuits and Systems*, pp 412–415.

(14) W. H. Jenkins and B. J. Leon. 1977. The use of residue number systems in the design of finite impulse response filters. *IEEE Transactions on Circuits and Systems*, CAS-24(4):191–201.

(15) W. Lynch et al. 1998. "Low load latency through sum-addressed memory". In: *Proceedings, 25th International Symposium on Computer Architecture*, pp 369–379.

(16) R. Zimmerman. 1999 "Efficient VLSI implementation of modulo $2^n \pm 1$ addition and multiplication". In: *Proceedings, 14th Symposium on Computer Arithmetic*, pp. 158–167.

(17) A. H. Hiasat. 2000. RNS arithmetic for large and small moduli. *IEEE Transactions on Circuits and Systems–II*, 47(9):937–940.

(18) M. Dugdale. 1994. Residue multipliers using factored decomposition. *IEEE Transactions on Circuits and Systems–II*, 41:623–627.

(19) D. Radhakrishnan and Y.Yuan. 1992. Novel approaches to the design of VLSI RNS multipliers. *IEEE Transactions on Circuits and Systems – II*, 39:52–57.

(20) G. A. Jullien. 1980. Implementation of multiplication modulo a prime number, with applications to number theoretic transforms. *IEEE Transactions on Computers*, C-29:899–905.

(21) D. Radhakrishnan and Y. Yuan. 1991. Fast and highly compact RNS multipliers. *International Journal of Electronics*, 70(2):281–293.

Chapter 6

Comparison, overflow-detection, sign-determination, scaling, and division

After the last long two chapters, the reader will probably be grateful for the slightness of this chapter. That slightness come about because, the operations discussed in this chapter being rather problematic ones, most work in RNS has been on applications that do not require these operations. The fundamental operation here is magnitude-comparison: overflow-detection and sign-determination are easily reduced to comparison, division requires comparisons together with additions/subtractions or multiplications, and scaling is division by a constant.

It should be noted that even RNS are restricted to applications in which the predominant operations are additions and multiplications, it may not be possible to completely do away with the problematic operations. In particular, scaling is especially important: In many of the algorithms for which RNS are particularly good, the most significant operation is the computation of inner products, which computation consists of sequences of multiply-and-add operations. It is therefore inevitable that in such processing the word-lengths required to accommodate intermediate results will grow, and, unless proper action is taken, overflow can occur. Values therefore need to be scaled to ensure that all computed results lie within the proper dynamic range; for normal RNS, this means ensuring that a computed result is the true result and not that result modulo the upper value of the dynamic range. Such scaling may be applied to inputs (e.g. in Number Theoretic Transforms) or to intermediate results (e.g. in Fast Fourier Transforms, Finite Impulse Response filters, etc.)

The chapter consists of a section on comparison, both exact and approximate, scaling (by base extension and through the use of the Core Function), and division (subtractive and multiplicative).

6.1 Comparison

A straightforward way to compare two residue numbers is to reverse-convert to a weighted representation and then carry out a conventional comparison. This, however, is costly. Three types of techniques are discussed in what follows. The first is approximate comparison, which is adequate for certain types of division algorithms and which is therefore discussed in more detail in Section 6.2. The second is the use of base extension, the basis of many methods that have been proposed for exact comparison. And the third is the use of the Core and Parity functions. In either case, there is getting around the fact that some magnitude-information must be extracted from the residue representations.

The Chinese Remainder Theorem (CRT), which relates a number, X, and its RNS representation, $\langle x_1, x_2, \ldots, x_N \rangle$, relative to moduli $m_1, m_2, \ldots, m_N$—that is, $x_i = |X|_{m_i}$—is

$$X = \left| \sum_{i=1}^{N} w_i x_i \right|_M$$

where $M = \prod_{i=1}^{N} m_i$ and $w_i = M_i |M_i^{-1}|_{m_i}$, with $M_i = M/m_i$ and $|M_i^{-1}|_{m_i}$ being the multiplicative inverse of M_i with respect to m_i. An "approximate" CRT may be obtained by dividing through by a constant. If we divide through by M, then we have

$$\frac{X}{M} = \left| \sum_{i=1}^{N} \frac{|M_i^{-1}|_{m_i}}{m_i} x_i \right|_1 \quad (6.1)$$

The values of each summand will be in the interval $[0, 1)$, and the final result after the addition is obtained by discarding the integer part and retaining the fractional part of the sum. The exact computation of Equation 6.1 is equivalent to reverse conversion and is likely to be quite costly in this context. But, provided enough accuracy is used, the results of such scaling may be used for comparison in cases where an exact comparison is not required.

EXAMPLE. Let $X = \langle 0, 6, 3, 0 \rangle$, $Y = \langle 5, 3, 0, 0 \rangle$ in the moduli-set $\{8, 7, 5, 3\}$. Then

$$\frac{X}{M} \approx |0.0000 + 0.8571 + 0.2000 + 0.0000|_1 \approx 0.0571$$

$$\frac{Y}{M} \approx |0.6250 + 0.4286 + 0.0000 + 0.0000|_1 \approx 0.0536$$

and we may conclude that $X > Y$ if the difference between the two is below the maximum possible difference-error. END EXAMPLE

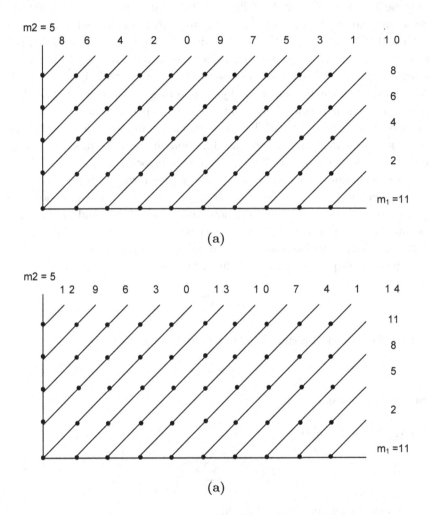

Figure 6.1: Ordering of numbers for comparison

6.1.1 Sum-of-quotients technique

The following method for exact comparison is given in [16]. The underlying idea is that in the finite N-dimensional space defined by the N moduli of an RNS, the integers may be ordered along lines parallel to the diagonal. In

Mixed-Radix Conversion, each line corresponds to the leading mixed-radix digit; an example is shown in Figure 6.1(a), for $N = 2$. The diagonals may be labelled different by following the natural order of the integers. So, corresponding to the example of Figure 6.1(a), we instead have Figure 6.1(b): 0, 1, 2, 3, and 5 label the first diagonal; 5, 6, 7, 8, and 9 label the second diagonal, 10 labels the third diagonal; 11, 12, 13, and 14 label the fourth diagonal; and so forth. (The new ordering is shown in Figure 6.1(b)). Numbers can then be compared by considering the diagonals that they belong to, which is essentially what is done when numbers are compared via Mixed-Radix Conversion. The essence of the sum-of-quotients technique is to devise a monotonicaly growing function that can be used to quickly determine the diagonal to which a given number belongs.

In what follows, we shall assume that we have the RNS is based on a set of pairwise relatively prime moduli, $m_1, m_2, \ldots, m_N$. M will denote the product of the moduli, M_i will denote the value M/m_i, and $\left|x^{-1}\right|_m$ will denote the multiplicative inverse of x with respect to the modulus m. We will also assume residue representations $X \cong \langle x_1, x_2, \ldots, x_N \rangle$

The sum-of-quotients, S_Q, is defined by

$$S_Q = \sum_{i=1}^{N} M_i$$

For the fundamental result, we also require the values s_i:

$$s_i = \left|-\left|m_i^{-1}\right|_{S_Q}\right|_{S_Q}$$

and we then have

THEOREM. The function

$$D(X) = \left|\sum_{i=1}^{N} s_i x_i\right|_{S_Q}$$

is a monotonically increasing function.

Given a number X, $D(X)$ is the diagonal on which it lies. Furthermore, we have this additional result

THEOREM. If $X < Y$ and $D(X) = D(Y)$, then $x_i < y_i$.

An example [17]:

EXAMPLE. Suppose we have $m_1 = 5, m_2 = 11, m_3 = 14, m_4 = 17$, and $m_5 = 9$. Then $S_Q = 62707$. Now, let $X = 30013 \cong \langle 3, 5, 11, 8, 7 \rangle$ and $Y = 11000 \cong \langle 0, 0, 10, 1, 2 \rangle$. Then $s_1 = 37624, s_2 = 45605, s_3 = 4479, s_4 = 51641, s_5 = 48772$. From which we get $D(X) = 15972$ and $D(Y) = 5854$, and $D(X) > D(Y)$, as we would expect. END EXAMPLE

In summary, for comparison of two RNS numbers, X and Ym with the sum-of-quotients technique, we first compute $D(X)$ and $D(Y)$. Then

- if $D(X) < D(Y)$, conclude that $X < Y$;
- if $D(X) > D(Y)$, conclude that $X > Y$;
- and if $D(X) = D(Y)$, check the individual residue-pairs to determine whether $X < Y$, $X = Y$, or $X >$.

6.1.2 Core Function and parity

The Core Function and concept of parity in RNS have been introduced in Chapter 2. Briefly, in a system with moduli $m_1, m_2, \ldots m_N$, (i.e. the dynamic range is $[0, M)$, where $M = \prod_{i=1}^{N} m_i$) the core, C_n, of a number n is given by

$$C_n = \sum_{i=1}^{N} w_i \left\lfloor \frac{n}{m_i} \right\rfloor$$

for certain weights fixed w_i. And the parity of a number is its residue with respect to the redundant modulus 2. Given $\langle x_1, x_2, \ldots, x_N \rangle$ as the RNS representation of a number X, $|X|_2$ can be found by base extension or computed during forward conversion and carried along. A parity of 0 indicates an *even number*, and 1 indicates an *odd number* [11].

Suppose that we are given two RNS numbers, $X \stackrel{\triangle}{=} \langle x_1, x_2, \ldots, x_N \rangle$ and $Y \stackrel{\triangle}{=} \langle y_1, y_2, \ldots, y_N \rangle$, with both corresponding represented numbers, $\widetilde{X}$ and $\widetilde{Y}$, in the range $[0, M)$, suppose that M is odd, and also suppose that (see Chapter 1) the interval $[0, M/2]$ represents positive numbers and the interval $(M/2, M)$ represents negative numbers. Then the following results hold and may be used for overflow-detection and sign-determination.

THEOREM. If C_M is odd, then $X + Y$ overflows if

- $X + Y$ is odd and $\widetilde{X}$ and $\widetilde{Y}$ have the same parity; or
- $X + Y$ is even and $\widetilde{X}$ and $\widetilde{Y}$ have different parities.

THEOREM. $\widetilde{X}$ is positive if and only if $\langle |2x_1|_{m_1}, |2x_2|_{m_2}, \ldots, |2x_N|_{m_N} \rangle$ is even.

As an alternative to the use of the Core Function, [10] has proved the following result, for odd M and assuming that the entire range is used for positive numbers.

THEOREM. If X and Y have same parity and $Z = X - Y$, then $X \geq Y$ iff Z is an even number. And $X < Y$ iff Z is an odd number.

THEOREM. If X and Y have different parities and $Z = X - Y$, then $X \geq Y$ iff Z is an odd number. And $X < Y$ iff Z is an even number.

6.2 Scaling

A straightforward way to perform scaling is to proceeds as follows. Suppose we have an RNS with moduli $m_1, m_2, \ldots, m_N$, and that $X \cong \langle x_1, x_2, \ldots, x_N \rangle$ (where $x_i = |X|_{m_i}$) is to scaled by K, to yield $Y \cong \langle y_1, y_2, \ldots, y_N \rangle$ (where $y_i = |Y|_{m_i}$). Then, by definition

$$X = YK + |X|_K$$

and, therefore,

$$Y = \frac{(X - |X|_K)}{K}$$

If $|K^{-1}|_{m_i}$, the multiplicative inverses of K with respect to the moduli m_i, exist then Y may be computed as

$$\begin{aligned} y_i &= |Y|_{m_i} \\ &= \left| |X - |X|_K|_{m_i} \times |K^{-1}|_{m_i} \right|_{m_i} \end{aligned}$$

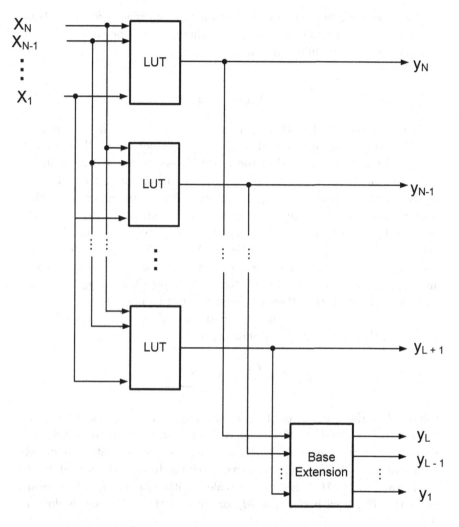

Figure 6.2: Architecture for scaling by base extension

So the scaling boils down to a set of additions (i.e. subtractions) and multiplications. We then need to compute $|X - |X|_K|_{m_i}$, and

$$|X - |X|_K|_{m_i} = |X|_{m_i} - ||X|_K|_{m_i}|_{m_i}$$
$$= |x_i - ||X|_{Km_i}|_{m_i}$$

The ease with which this computation can be carried out depends on how difficult it is to compute $|X|_K$ and, therefore, on the choice of K. If K is the smallest modulus, m_j, then

$$|x_i - |X|_K|_{m_i} = |x_i - x_j|$$

If K is a product of a subset, $m_1, m_1, \ldots, m_L$, of the moduli, then $|X|_K$ will be easy to compute; but then $|K^{-1}|_{m_i}$ will not exist. Nevertheless, that need not signal the end of the world: Observe that Y will be in the dynamic range, since $0 \leq Y < \prod_{i=L+1}^{N} m_i$. Therefore, the result of the scaling is completely defined by $\langle y_1, y_2, \ldots, y_N \rangle$. And if we can compute this, then we also can, by base extension, compute $\langle y_1, y_2, \ldots, y_L \rangle$. Furthermore, it is evident that $\langle y_1, y_2, \ldots, y_N \rangle$ is completely determined by the values of $x_{L+1}, x_{L+2}, \ldots, x_N$. So the scaling may be implemented as follows. $x_{L+1}, x_{L+2}, \ldots, x_N$ are used to address lookup tables whose outputs are $y_1, y_2, \ldots, y_L$. Base extension is then used to obtain $y_{L+1}, y_{L+2}, \ldots, y_N$. The basic architecture is therefore as shown in Figure 6.1 [1].

The Core Function can also be used for scaling, as follows [12]. By definition, the core, $C(X)$, of a number, X, is

$$C(X) = \frac{C(M)}{M} X - \sum_{i=1}^{N} \frac{w_i}{m_i} x_i$$

where M is the product of the moduli, $m_1, m_2, \ldots, m_N$, employed, and $x_i = |X|_{m_i}$. Therefore, if we can compute cores within the RNS, then we can approximately scale X. If the moduli-set is split into two subsets such that M_J and M_K are, respectively, the products of the moduli in the subsets and $M_J \approx M_K$, then we can scale by either M_J or M_K. This means extracting $C(X)$ with $C(M) = M_J$ or with $C(M) = M_K$ and is done as follows.

The Chinese Remainder Theorem for Core functions is (Chapter 7)

$$|C(X)|_{C(M)} = \left| \sum_{i=1}^{N} x_i C(X_i) \right|_{C(M)}$$

where

$$X_i = \frac{M}{m_i} \left| \left(\frac{M}{m_i} \right)^{-1} \right|_{m_i}$$

and $(M/m_i)^{-1}|_{m_i}$ is the multiplicative inverse of M/m_i with respect to m_i. Extracting the core with respect to M_J, i.e. setting $C_J(M) = M_J$, we have

$$|C_J(X)|_{C_J(M)} = \left| \sum_{j=1} x_j C(X_j) \right|_{C_J(M)}$$

Since m_j is a factor of $C_J(M)$, we have

$$\left| |C_J(X)|_{C_J(M)} \right|_{m_i} = |C_J(X)|_{m_j}$$

$$= \left| \sum_{j=1} x_i C(X_i) \right|_{m_j}$$

This does not apply to the moduli that make up M_K. Similarly for the moduli, m_k, that make up M_K, but not those that make up M_J, we have

$$|C_K(X)|_{m_k} = \left| \sum_{i=1} x_i C(X_i) \right|_{m_k}$$

So with the moduli that make up M_J, we can compute $C_J(X) \approx X/M_K$, and with the moduli that make up M_K, we can compute $C_J(X) \approx X/M_J$. To obtain scaled residues across all moduli, we first compute $\Delta C(X) \triangleq C_J(X) - C_K(X)$ and then add to or subtract that value from the values from one subset. $\Delta C(X)$ is computed as

$$|\Delta C(X)|_{\Delta C(M)} = \left| \sum_i \Delta C(X_i) \right|_{\Delta C(M)}$$

6.3 Division

Conventional algorithms for division have been introduced in Chapter 5 and may be classified as either *subtractive* (i.e. the primary operation is subtraction) or *multiplicative* (i.e. the primary operation is multiplication). We shall consider how both types can be adapted for RNS division.

6.3.1 Subtractive division

6.3.1.1 Basic subtractive division

Basic mechanical division of two integers consists of a sequence of magnitude-comparisons and shift-and-subtract operations, along with some implicit or explicit magnitude-comparisons. For the division of X ($2n$ bits) by Y (n bits), where, without loss of generality, we shall assume that both operands are positive, the basic binary algorithm is

(1) Set i to 0, $X^{(i)}$, the ith partial remainder, to X, and $Q^{(i)}$, the ith partial quotient, to 0.
(2) Select a quotient digit, q_i, and append it to the quotient as formed so far: $Q^{(i+1)} = 2Q^{(i)} + q_i$.
(3) Reduce the partial remainder: $X^{(i+1)} = 2X^{(i)} - q_i Y$.
(4) Increment i; if $i \neq n$, goto (2).

At the end of the process, $Q^{(n)}$ will be the quotient, and $X^{(n)}$ will be the remainder. Note that here we are shifting the partial quotient to the left and appending the new quotient digit. (And a similar remark applies to the reduction of the remainder.) This is simply a matter of convenience: we could equally well hold the remainder in place and at each step shift down the quotient-digit, i.e. add q_i scaled down by a power of two and reduce the partial remainder by the product of the quotient-digit and the divisor.

The process just described is the basic *subtractive-division* algorithm, of which there are a few variants. The most common of such variants are the *non-restoring* and SRT algorithms, both of which have been described in Section 5.2.1. There is also the *restoring* algorithm, which differs from the non-restoring one in that an "unsuccessful" subtraction, i.e. one that, from an incorrect guess, leaves the partial remainder negative, is immediately followed by a subtraction to restore its value, and another guess is then made. In general, subtractive-division algorithms differ in how a guess is made for the next quotient-digit, q_i, and in how a correction is made if the guess turns out to have been wrong.

The SRT algorithms are generally the fastest subtractive-division algorithms, because they rely on redundant signed-digit representations, and these allow more tolerance in how wrong a guess may be. Nevertheless, the basic restoring-division algorithm can be speeded up in a variety of ways. One of those ways is to shift over 0s in the partial remainder without performing any arithmetic [6]. The rationale for this is that if the divisor has been "normalized" so that its most significant bit is not 0, then as long as the leading bit of the partial remainder is 0, no successful subtraction can take place; so the quotient digit corresponding to each leading 0 in the partial remainder is immediately known to be 0. In non-restoring division, since the partial remainder can be positive or negative, it is possible to shift over both 0s and 1s. With current technology, shifting by arbitrary distances over 0s or 1s is no longer an effective way to realize high-speed division, and the SRT algorithm is preferred. Nevertheless, the basic idea of restoring division with shifting-over-0s has been put to good use in RNS division [7]. We next describe this.

The main problem with a restoring-division algorithm is that extra arithmetic operations are sometimes required to restore the partial remainder. The non-restoring algorithm avoids this, but, evidently, if any way could be found to ensure that the partial remainder is always positive, then these extra operations can be eliminated. In the algorithm to be described that is done by selecting q_i in such a way that q_iY is always smaller than the partial remainder. Also note that instead of initially normalizing the divisor and repeatedly shifting over 0s in partial remainder while entering an equal number of 0s in the quotient, we could just as well determine the number of 0s to entered in the quotient by repeatedly comparing the position of the leading non-0 bit in the partial remainder with the leading non-0 bit in the divisor: if the former is j and the latter is k, then the number of 0s to be entered in the quotient is $j - k$. The algorithm below uses $j - k - 1$ instead, thus ensuring that the partial remainder is always positive. That is, in binary the multiple of the divisor that is subtracted in Step (3) of the algorithm is either 0 or $2^{j-k-1}Y$. Based on these observations, the corresponding algorithm for RNS division is as follows.

Let the two RNS operands be $X \triangleq \langle x_1, x_2, \ldots, x_N \rangle$ and $Y \triangleq \langle y_1, y_2, \ldots, y_N \rangle$, where X is the the dividend and Y is the divisor. (Without loss of generality, we shall assume that both represent positive numbers.) And suppose X represents the number $\widetilde{X}$ and Y represents the number $\widetilde{Y}$, i.e. $X \cong \widetilde{X}$ and $Y \cong \widetilde{Y}$. Then the RNS restoring-division algorithm, with a "shifting over 0s", is

(1) Set i to 0 and $X^{(i)}$, the ith partial remainder, to X, and $Q^{(i)}$, the ith partial quotient, to 0.
(2) Find the position, k, of the most-significant non-0 bit of $\widetilde{Y}$. (This is equivalent to normalizing $\widetilde{Y}$.)
(3) Find the position, j, of the most-significant non-0 bit of $\widetilde{X}$. (This is equivalent to normalizing $\widetilde{X}$.)
(4) If $j > k$ (i.e. a successful subtraction must occur), then
   (a) set $Q^{(i+1)} = Q^{(i)} + q_i$, where $q_i = 2^{j-k-1}$;
   (b) set $X^{(i+1)} = X^{(i)} - q_iY$;
   (c) increment i and goto (3) if $i \neq n$.
(5) If $j = k$ (i.e. a subtraction may or may not be successful), then

(a) set $X^{(i+1)} = X^{(i)} - Y$;
(b) find the position, l, of the leading non-0 bit of $|X^{(i+1)}|$;
(c) if $l < j$, then set $Q^{(i+1)} = Q^{(i)} + q_i$.

This algorithm is equally valid for both integer and fractional representations.

All of the arithmetic operations in the preceding algorithm are in the RNS domain. Nevertheless, at Steps (2) and (3), we do not have $\tilde{X}$ and $\tilde{Y}$ but must extract some magnitude-information, partial though it may be. There are evidently a variety of ways in which this could be done. [7] proposes the use of lookup tables that are addressed by residues and whose outputs are the sought indices. That work proposes two architectures. The first assumes integer representations, exactly as given in the algorithm, and is suitable for a small or medium dynamic range. It has the structure shown in Figure 6.3. The values j and k are obtained from a lookup table and then used to address another lookup table that contains the values of q_i. A multiplier then computes $q_i Y$, the value of which is subsequently subtracted from the current partial remainder, to form a new partial remainder. Concurrently, the partial quotient is updated.

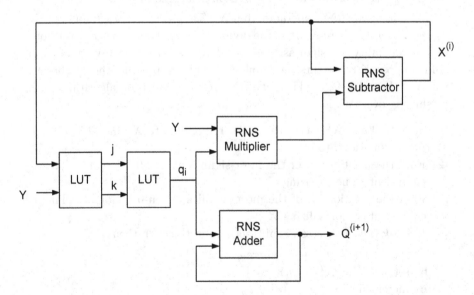

Figure 6.3: Architecture for subtractive division (small range)

For a large dynamic range, the lookup tables would be prohibitively large if integer representations are assumed. Therefore, in this case fractional representations are assumed. These representations are obtained by scaling by M, as in the approximate Chinese Remainder Theorem (Equation 6.1). The architecture is shown in Figure 6.4. The residues of X are used to address lookup tables that hold the CRT-values $(|x_i|M_i^{-1}||_{m_i}|)/m_i$; these are then added, modulo 1, to obtain X/M. Instead of using another lookup table, as above, priority encoders are used to obtain j and k. The rest of the architecture is similar to that in Figure 6.3.

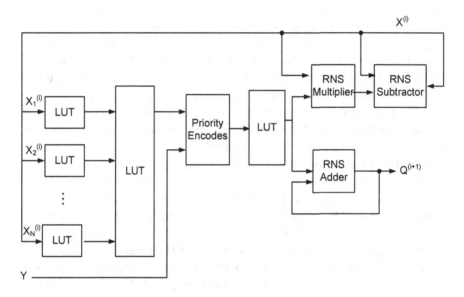

Figure 6.4: Architecture for subtractive division (large range)

The algorithms and architectures above have been obtained by modifying a conventional restoring algorithm, so that restoration of partial remainders is not required, and then shifting over 0s in the partial remainder. For conventional division, more sophisticated and faster algorithms—based on both restoring and non-restoring algorithms, shifting over both 0s and 1s, using several multiples of the divisor, employing radices larger than two—are described in [6]. It is conceivable that these too could be adapted for RNS division.

6.3.1.2 Pseudo-SRT division

We now describe an "SRT" division algorithm [3]; the real SRT algorithm is described in Section 5.1. Let $F(X) = X/M$ (Equation 6.1). Even for unsigned operands, intermediate results in SRT will be signed. For normal RNS, signed numbers are accommodated by dividing the dynamic range into two halves (Chapter 1). Here, $F(X) \in [0,1)$; so the range $[0, 1/2)$ corresponds to the positive numbers, and $[1/2, 1)$ corresponds to the negative numbers. In the former case, the magnitude of the number is $F(X)$, and in the latter case the magnitude is $1 - F(X)$. As noted above, computing $F(X)$ with sufficient accuracy for an exact comparison is too costly. But for pseudo-SRT division, approximations are sufficient. Let $F_a(X)$ denote the computed approximation to $F(X)$. Then, with regard to sign-determination, we have three possible results: X is definitely positive, X is definitely negative, and X is of indeterminate sign. We shall use $S_a(X)$ to denote the approximate sign of X; that is, $S_a(X) \in [+, -, +/-]$. The indeterminate case means that, because of limited accuracy, there are bands around 0, 1, and 1/2 such that numbers in the bands could be positive or negative, but we are unable to tell which. If we exclude the band around 1/2, then an indeterminate sign means that X is near 0 or near 1, i.e. it is of small magnitude. Such exclusion means that part of the dynamic range cannot be used, but, by appropriate choice, that part can be made negligibly small. The exclusion is done by ensuring that the *input* number is the range $[-(1/2 - 2^{-a})M, (1/2 - 2^{-a})M]$, for some positive integer a. Then, if $S_a(X) = +/1$, X is guaranteed to be in the range $[-2^{-a}M, 2^{-a}M]$.

$F_a(X)$ is computed as follows. A lookup table is constructed for each modulus m_i. In each such table, the jth entry ($j = 0, 2, \ldots, m_i - 1$) is the value

$$F_a(i,j) \triangleq \left\lfloor \left| \frac{j}{m_i} |M_i^{-1}|_{m_i} \right|_1 \right\rfloor$$

computed to b bits, where $b = a + \log N$ and a is chosen for the constraints above. $F_a(X)$ is then computed as

$$F_a(X) = \left| \sum_{i=1}^{N} F_a(i,j) \right|_1$$

And $S_a(X)$ is computed as

- $+$ (and $X \geq 0$) if $0 \leq F_a(X) \leq \frac{1}{2}$,
- $-$ (and $X < 0$) if $\frac{1}{2} \leq F_a(X) < 1 - 2^{-a}$,

- $+/-$ otherwise.

On the basis of the above, given two positive numbers, N and D, the following algorithm computes Q and R such that $N = QD + R$, with $0 \le R < D$.

(1) Set $j = 0$ and $Q = 0$
(2) While $S_a(\lfloor M/8 \rfloor - 2D) \ne -$ do $\{D \Leftarrow 2 \times D, j \Leftarrow j + 1\}$
(3) While $S_a(N - D) \ne -$ do $\{N \Leftarrow N - 2 \times D, Q \Leftarrow Q + 2\}$
(4) For $i = 1, 2, 3, \ldots, j$ do
   - if $S_a(N) = +$, then $\{N \Leftarrow 2 \times (N - D), Q \Leftarrow 2 \times (Q + 1)\}$
   - if $S_a(N) = -$, then $\{N \Leftarrow 2 \times (N + D), Q \Leftarrow 2 \times (Q - 1)\}$
   - if $S_a(N) = +/-$, then $\{N \Leftarrow 2 \times N - D, Q \Leftarrow 2 \times\}$
(5) if $S_a(N) = +$, then $\{N \Leftarrow N - D, Q \Leftarrow Q + 1\}$
(6) if $S_a(N) = +$, then $\{N \Leftarrow N - D, Q \Leftarrow Q + 1\}$
(7) if $S_a(N) = -$ or $(S_a(N) = +/-$ and $S(N) = -)$, then $\{N \Leftarrow N + D, Q \Leftarrow Q - 1\}$
(8) $R \Leftarrow 2^{-j}N$

The similarities between this algorithm and the conventional SRT algorithm are evident. First, the dividend and the divisor are normalized. Then, in the main loop, the quotient is repeatedly adjusted by -1, or 0, or 1, and the partial remainder accordingly adjusted by $-D$, or 0, or $+D$; this corresponds to the radix-2 SRT algorithm, i.e. the digit-set is $\{\bar{1}, 0, 1\}$. Lastly, a correction of -1 may be required of the quotient. And the final remainder is produced by scaling down the partial remainder.

For this algorithm, $a = 4$; so $2^{-a}M = M/16$. The dividend can be in the full range, i.e. $[0, M/2)$, but the divisor must not be larger than $3M/16$. In line 7, exact sign-determination (i.e. comparison against 0) is required, but it is the only such operation. The task of scaling by 2^j can be simplified by using only odd moduli; then all the residues of 2^j can be precomputed and stored. If one of moduli is even, then the scaling can be performed by base extension. A faster modification of the above algorithm is given in [3].

6.3.2 Multiplicative division

Conventional multiplicative-division algorithms too can be adapted for RNS, as has been done for the subtractive-division algorithms. The most common algorithm for multiplicative division is that based on the Newton-Raphson procedure (Section 5.2.2), in which one computes the reciprocal of the divisor and then multiplies that with the dividend to get the quotient. We next describe one such algorithm, from [9].

In order to be able to represent the large intermediate results that occur from multiplications, the arithmetic is implemented in an "extended" RNS whose range is approximately the square of the range of the "base" RNS. If the base RNS has N moduli, then the extended RNS has $2N$ moduli, $m'_1, m'_2, \ldots, m'_{2N}$, where $1 < m'_1 < m'_2 < \ldots < m'_{2N}$. These moduli are partitioned into two sets: $\{m_1 = m'_1, m_2 = m'_3, \ldots, m_N = m'_{2N-1}\}$ and $\{m_{N+1} = m'_2, m_{N+2} = m'_4, \ldots, m_{2N} = m'_{2N}\}$

Let M be $\prod_{i=1}^{N} m_i$ and $\widetilde{M}$ be $\prod_{i=N+1}^{2N} m_i$. Then it is the case that (a) $|M^{-1}|_{\widetilde{M}}$, the multiplicative inverse of M, with respect to $\widetilde{M}$, exists, since M and $\widetilde{M}$ are relatively prime; (b) for large N and moduli of similar magnitude, $\widetilde{M} - M$ will be small; and (c) if X and Y are in the base range, then, because $M^2 < M\widetilde{M}$, the multiplication $X \times Y$ will never overflow.

Without initially considering how the arithmetic operations are to be carried out in RNS, the quotient (Q) and the remainder (R) from the integer division of X (where $0 \leq X < M$) by Y (where $1 \leq Y < M$) are obtained by the algorithm

(1) Set Q to $\lfloor X \times \lfloor M/Y \rfloor / M \rfloor$.
(2) Set R to $X - Q \times Y$.
(3) If $R \geq Y$, then
   (a) increment Q;
   (b) subtract Y from R.

We therefore need to compute a reciprocal, $\lfloor M/Y \rfloor /$ and then scale that by M. The first is carried out by an application of the Newton-Raphson root-finding procedure:

$$Z_{i+1} = Z_i - \frac{f(Z_i)}{f'(Z_i)}$$

For the computation of M/Y, $f(Z_i) = M/Z_i - Y$, whence the recurrence

$$Z_{i+1} = \frac{Z_i(2M - YZ_i)}{M}$$

Since we require only integer values, this recurrence becomes

$$Z_{i+1} = \left\lfloor \frac{Z_i(2M - YZ_i)}{M} \right\rfloor$$

And the algorithm to compute the reciprocal $\lfloor M/Y \rfloor$ is

(1) Set Z_1 to 0 and Z_2 to 2.

(2) If $Z_i \neq Z_{i+1}$, then

    (a) set Z_i to Z_{i+1};

    (b) set Z_{i+1} to $\lfloor Z_i \times (2M - YZ_i)/M \rfloor$.

(3) Goto (2).

(4) If $M - Y \times Z_{i+1} < Y$, then the result is Z_{i+1}; otherwise the result is Z_{i+1}.

The last step is a correction step that is required to deal with cases such as $M = 10, Y = 3, \lfloor M/Y \rfloor = 3, Z_i = Z_{i+1} = 2$.

In the algorithm as given above, the starting value Z_2 has been arbitrarily set to 2. Now, the number of iterations required of the Newton-Raphson procedure depends on how accurate the starting value is; so the execution of the algorithm can be speeded up by setting Z_2 to a better approximation of $\lfloor M/Y \rfloor$. This may be done by, for example, using a lookup table that stores the approximations and is addressed with some part of the representation of Y. Many approximation techniques (for the conventional Newton-Rapshon) are given in [6], and these can be adapted for the case at hand. We next turn to the RNS implementation of the arithmetic operations. We shall assume that the original operands, X and Y, have been base-extended into the extended RNS.

The additions, subtractions, multiplications, and test for equality are easily implemented in the RNS domain: simply apply the operation to residue pairs. (In the last case, $X = Y$, if and only if $x_i = y_i$, for all i, where $x_i = |X|_{m_i}$ and $y_i = |Y|_{m_i}$.) Scaling by M is required in both the main algorithm and the Newton-Raphson algorithm. That scaling may be done in one of two different ways. The first is as follows. Let $U \cong \langle u_1, u_2, \ldots, u_{2N} \rangle$ be the number to be scaled. Then $\langle u_1, u_2, \ldots, u_N \rangle$ represents the remainder $R = |U|_M$ and $\langle u_{N+1}, u_{N+2}, \ldots, u_{2N} \rangle$ represents MQ. The latter is base-extended into the extended RNS and the result subtracted from M. Multiplying the result of the subtraction by $|M^{-1}|_{\widetilde{M}}$ yields the quotient, $Q = \langle q_{N+1}, q_{N+2}, \ldots, q_{2N} \rangle$, which is then base-extended into the extended RNS, i.e. into $\langle q_1, q_2, \ldots, q_N, q_{N+1}, q_{N+2}, \ldots, q_{2N} \rangle$. The second way to do the scaling is through Mixed-Radix Conversion. Suppose the mixed-radix representation of U is $(v_1, v_2, \ldots, v_{2N})$. Then, by Equation 7.9,

$$U = \sum_{i=1}^{2N} v_i \prod_{j=1}^{i-1} m_i$$

$$= \sum_{i=1}^{N} v_i \prod_{j=1}^{i-1} m_i + M \sum_{i=N+1}^{2N} v_i \prod_{j=N+1}^{i-1} m_i$$
$$\triangleq R + MQ$$

with $R = |U|_M$ and $Q = \lfloor U/M \rfloor$. Q is the sought result and is converted back into the extended RNS by evaluating the summation for each modulus. Lastly, we have the difficult operation of comparison. Two methods for RNS implementation are given in [9], and we have discussed others above.

We leave it as an exercise for the reader to devise an architecture for the implementation of the algorithm.

6.4 Summary

The arithmetic operations discussed here are the most problematic in RNS. Many applications of RNS are in areas in which these operations are rarely required, but even so they cannot be completely avoided; as an example, scaling may be required to ensure that intermediate results from multiplication stay in range. With regard to division, there appears to be a great deal of scope for the development of new algorithms based on those for conventional division. We have not discussed square-root, but given that conventional algorithms for square-root are very similar to conventional algorithms for division, it should not be too hard to devise square-root algorithms. For example, an examination of the conventional "SRT"-type of square-root algorithm show that the RNS "SRT" division-algorithm above can easily be modified to obtain an algorithm for square-roots.

References

(1) A. Garcia. 1999. A lookup scheme for scaling in the RNS. *IEEE Transactions on Computers*, 48(7):748–751.
(2) C. Y. Hung and B. Parhami. 1995. Error analysis of approximate Chinese Remainder Theorem decoding. *IEEE Transactions on Computers*, 44(11):1344–1349.
(3) C. Y. Hung and B. Parhami. 1994. An approximate sign detection method for residue numbers and its application to RNS division. *Computers and Mathematics with Applications*, 27(4):23–35.
(4) T. V. Vu. 1985. Efficient implementation of the Chinese Remainder

Theorem for sign detection and residue decoding. *IEEE Transactions on Computers*, 34:645–651.

(5) G. A. Jullien. 1978. Residue number scaling and other operations using ROM arrays. *IEEE Transactions on Computers*, 27(4):325–337.

(6) A. R. Omondi. 1994. *Computer Arithmetic Systems*. Prentice-Hall, UK.

(7) A. Hiasat and H. Abdel-Aty-Zohdy. 1999. Semicustom VLSI design of a new efficient RNS division algorithm. *The Computer Journal*, 42(3):232–240.

(8) E. Kinoshita, H. Kosako, and Y. Kojima. 1973. General division in the symmetric residue number system. *IEEE Transactions on Computers*, C-22:134–142.

(9) M. A. Hitz and E. Kaltofen. 1995. Integer division in residue number systems. *IEEE Transactions on Computers*, 44(8):983–989.

(10) M. Lu and J.-S. Chiang. 1992. A novel division algorithm for the residue number system. *IEEE Transactions on Computers*, 41(8):1026–1032.

(11) D.D. Miller. 1986. Analysis of the residue class core function of Akushskii, Burcev, and Park. In: G. Jullien, Ed., *RNS Arithmetic: Modern Applications in Digital Signal Processing*. IEEE Press.

(12) N. Burgess. 2003. Scaling an RNS number using the core function. In: *Proceedings, 16th IEEE Symposium on Computer Arithmetic*.

(13) F. Barsi and M. C. Pinotti. 1995. Fast base extension and precise scaling in RNS for look-up table implementations. *IEEE Transactions on Signal Processing*, SP-43:2427–2430.

(14) A. Shenoy and R. Kumaresan. 1989. A fast and accurate RNS scaling technique for high speed signal processing. *IEEE Transactions on Acoustics Speech and Signal Processing*, ASSP-37:929–937.

(15) Z. D. Ulman and M. Czyak. 1988. Highly parallel, fast scaling of numbers in nonredundant residue arithmetic. *IEEE Transactions on Signal Processing*, SP-46:487–496.

(16) G. Dimauro, S. Impedovo, and G. Pirlo. 1993. A new technique for fast number comparison in the residue number system. *IEEE Transactions on Computers*, 42(5):608–612.

Chapter 7

Reverse conversion

Reverse conversion is the process, usually after some residue-arithmetic operations, of translating from residue representations back to conventional notations. It is one of the more difficult RNS operations and has been a major, if not *the* major, limiting factor to a wider use of residue number systems. The main methods for reverse conversion are based on the *Chinese Remainder Theorem* (CRT) and the *Mixed-Radix Conversion* (MRC) technique. All other methods are just variations of these two, which variations arise from either the type of moduli-set chosen or from certain properties that can be easily adapted to suit the particular approach chosen.

The chapter consists of three main sections, one each for the CRT, MRC, and the Core Function. Although the last of these is nominally different, close examination will reveal that it is in fact just a variation on the CRT theme; but the details of the formulation are such that they warrant a separate section. In addition to discussions of the basic theoretical foundations of the main methods, the chapter also includes some discussions on architectural implementations.

7.1 Chinese Remainder Theorem

Conversion from residue numbers to conventional equivalents seems relatively straightforward on the basis of the Chinese Remainder Theorem (CRT), introduced in Chapter 1. Unfortunately, the direct realization of an architectural implementation based on this theorem presents quite a few problems, and, compared to forward conversion, a generalized realization (i.e. for arbitrary moduli) is likely to be both complex and slow.

In what follows, we shall first derive (from first principles) the CRT and in doing so also show how it applies in reverse conversion. We shall do that with the aid of a small (5-moduli) example. Subsequently, we shall consider hardware architectures for the implementation of the CRT.

Let the residue representation of X be $\langle x_1, x_2, x_3, x_4, x_5 \rangle$ relative to the pair-wise relatively-prime moduli $m_1, m_2, \ldots, m_N$. That representation may be written as

$$X \cong \langle x_1, x_2, x_3, x_4, x_5 \rangle$$
$$= \langle x_1, 0, 0, 0, 0 \rangle + \langle 0, x_2, 0, 0, 0 \rangle + \langle 0, 0, x_3, 0, 0 \rangle + \langle 0, 0, 0, x_4, 0 \rangle$$
$$+ \langle 0, 0, 0, 0, x_5 \rangle$$
$$\stackrel{\triangle}{=} X_1 + X_2 + X_3 + X_4 + X_5$$

where X_i is a conventional number. Finding each X_i is also a reverse-conversion process, but it is one that is much easier process than that of finding X. And once the X_is are available, the number X can be obtained easily.

Consider now the problem of determining X_1 from $\langle x_1, 0, 0, 0, 0 \rangle$. Observe that X_1 is a multiple of m_2, m_3, m_4 and m_5, since the residue with respect to each of those moduli is zero. So X_1 may be expressed as

$$X_1 \cong x_1 \times \langle 1, 0, 0, 0, 0 \rangle$$
$$\stackrel{\triangle}{=} x_1 \widetilde{X_1}$$

where $\widetilde{X_1}$ is a conventional number. We therefore need to find a number $\widetilde{X_1}$ such that $\left|\widetilde{X_1}\right|_{m_1} = 1$. Now, recall (from Chapter 2) that multiplicative inverse, $|x^{-1}|_m$, of a number, x, with respect to a modulus m is defined by

$$\left||x^{-1}|_m \, x\right|_m = 1$$

Let M be the product $m_1 m_2 m_3 m_4 m_5$, and define M_i as M/m_i. Then, by definition

$$\left||M_1^{-1}|_{m_1} M_1\right|_{m_1} = 1$$

This inverse exists since the m_j are pair-wise relatively-prime. Therefore

$$\widetilde{X_1} = \left|M_1^{-1}\right|_{m_1} M_1$$

and

$$X_1 = x_1 \widetilde{X_1}$$
$$= x_1 \left|M_1^{-1}\right|_{m_1} M_1$$

The last equation can then be extended readily to the i-th modulus:

$$X_i = x_i \left|M_i^{-1}\right|_{m_i} M_i$$

The number X that we seek is then just the sum of X_is:

$$X = \sum_{i=1}^{5} X_i$$

$$= \sum_{i=1}^{5} x_i \left|M_i^{-1}\right|_{m_i} M_i$$

Here X is any number whose residue representation is $\langle x_1, x_2, x_3, x_4, x_5 \rangle$. It may be the number X that is so obtained is not within the dynamic range of the given residue number system. In such a case, a reduction modulo M will yield a number that is within the correct range; that is, what we will compute instead is

$$|X|_M = \left| \sum_{i=1}^{5} x_i \left|M_i^{-1}\right|_{m_i} M_i \right|_M$$

This last expression is essentially the CRT, and what we have done above is derive it first principles. The theorem shows, in a single expression, how a number may be obtained from its residues. The arithmetic requirements include several multiply-add operations followed by a modular reduction.

In sum, then, here is the statement of the Chinese Remainder Theorem. Given a set of pair-wise relatively-prime moduli, $m_1, m_2, \ldots, m_N$, and a residue representation $\langle x_1, x_2, \ldots, x_N \rangle$ in that system of some number X, i.e. $x_i = |X|_{m_i}$, that number and its residues are related by the equation

$$|X|_M = \left| \sum_{i=1}^{N} x_i \left|M_i^{-1}\right|_{m_i} M_i \right|_M \tag{7.1}$$

where M is the product of the m_is. If the values involved are constrained so that the final value of X is within the dynamic range, then the modular reduction on the left-hand side may be omitted.

Equation 7.1 shows clearly the primary difficulty in the application of the CRT: the need for the modular reduction of a potentially very large number relative to a large modulus (M).

EXAMPLE. Consider the moduli-set $\{3, 5, 7\}$, and suppose we wish to find the X whose residue representation is $\langle 1, 2, 3 \rangle$. To do so, we first determine the M_is and their inverses:

$$M_1 = M/m_1$$
$$= \frac{3 \times 5 \times 7}{3}$$
$$= 35$$

$$M_2 = M/m_2$$
$$= \frac{3 \times 5 \times 7}{5}$$
$$= 21$$

$$M_3 = M/m_3$$
$$= \frac{3 \times 5 \times 7}{7}$$
$$= 15$$

Whence

$$\left| M_1 M_1^{-1} \right|_3 = 1$$
$$\left| 35 M_1^{-1} \right|_3 = 1$$
$$M_1^{-1} = 2$$

Similarly

$$\left| M_2 M_2^{-1} \right|_5 = 1$$
$$\left| 21 M_2^{-1} \right|_5 = 1$$
$$M_2^{-1} = 1$$

and

$$\left| M_3 M_3^{-1} \right|_7 = 1$$
$$\left| 15 M_3^{-1} \right|_7 = 1$$
$$M_3^{-1} = 1$$

Then, by the CRT, we have ($M = 3 \times 5 \times 7 = 105$)

$$X = \left| \sum_{i=1}^{3} x_i X_i \right|_{105}$$

$$= |1 \times 35 \times 2 + 2 \times 21 \times 1 + 3 \times 15 \times 1|_{105}$$
$$= | \, |112|_{105} + 45|_{105}$$
$$= |7 + 45|_{105}$$
$$= 52$$

END EXAMPLE

A simpler way of looking at the conversion from residues is to consider the integer that corresponds to the residue representation that has a 1 in the ith residue position and 0s in all other residue positions. We shall refer to this integer is as the *weight* of the ith residue. The ordering of the residues, x_i, is of no consequence, since there will be only one integer whose residue representation has 0s in all positions not equal to i and a 1 in the i-th position. The conventional equivalent can therefore be determined by scaling the weighted sum of residues up to the integer representation modulo M. With this construction of weights w_i, we have

$$X = \left| \sum_{j} x_j w_j \right|_{M}$$

This is so since w_j is a multiple of all the m_i ($i \neq j$), and $|x_j w_j|_{m_j} = |X|_{m_j}$, for all j. In order to recover the sought integer from its residue representation, we need to sum the weighted residues modulo M. The process is illustrated in the next example.

EXAMPLE. Consider the moduli-set $\{2, 3, 5\}$. The range of numbers that can be represented in this set is $R = [0, 29]$. Let us consider a number X in this range, with the residue representation $\langle 1, 2, 4 \rangle$. The weights for set can be obtained as follows.

$$w_1 = y \times m_2 \times m_3$$
$$= y \times 3 \times 5$$

In the above we need to determine w_1 with $y = 1$:

$$w_1 = 1 \times 3 \times 5$$
$$= 15$$

Similarly, we may determine w_2 and w_3 as follows.

$$w_2 = m_1 \times y \times m_3$$
$$= 2 \times 1 \times 5$$
$$= 10$$

and

$$w_3 = m_1 \times m_2 \times y$$
$$= 2 \times 3 \times 1$$
$$= 6$$

The integer that corresponds to the above residue representation $\langle 1, 2, 4 \rangle$ is then

$$X = \left| \sum_j x_j w_j \right|_{30}$$
$$= |1 \times 15 + 2 \times 10 + 4 \times 6|_{30}$$
$$= 29$$

END EXAMPLE

A straightforward way to implement to implement Equation 7.1 is as follows. The constants $X_i \triangleq M_i \left| M_i^{-1} \right|_{m_i}$ are multiplied, in parallel, with the residues, x_i, and the results then added in a multi-operand modulo-M adder. Evidently, for a large dynamic range, large or many multipliers will be required, according to whether the range is obtained from a few, large moduli or from many, small moduli, and this may be costly. It should, however, be noted that the multipliers here need not be full multipliers: because one of each pair of operands is a constant, each multiplier can be optimized according to that particular operand. Depending on the actual technology used for the realization, i.e. on the relative cost and performance figures, an obvious way to avoid the use of multipliers is to replace them with look-up tables (say ROMs). That is, for each value of X_i, have a ROM that stores all the possible values of xX_i, for $x = 0, 1, 2, \ldots, m_i - 1$. Then

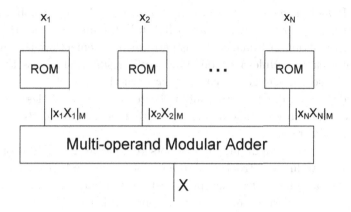

Figure 7.1: ROM-based CRT reverse converter

for each tuple, $\langle x_1, x_2, \ldots, x_N \rangle$, of residues the appropriate values are concurrently read out and added in a multi-operand modulo-M adder. The storage requirements or multipliers (in the case of combinational logic) can be kept down by rewriting applying a reformulated version of Equation:

$$\begin{aligned}
X &= \left| \sum_{i=1}^{N} x_i \left| M_i^{-1} \right|_{m_i} M_i \right|_M \\
&= \left| \sum_{i=1}^{N} |x_i|_{m_i} \left| M_i^{-1} \right|_{m_i} M_i \right|_M \\
&= \left| \sum_{i=1}^{N} \left| x_i \left| M_i^{-1} \right|_{m_i} \right|_{m_i} M_i \right|_M \\
&\stackrel{\triangle}{=} \left| \sum_{i=1}^{N} x_i X_i \right|_M
\end{aligned}$$

The resulting high-level architecture is then as shown in Figure 7.1.

With either of the general approach outlined above, the multi-operand modular adder may be realized as purely combinational-logic, or all-ROM (e.g. [3, 11]), or a mixture of both. Almost all architectures that have been proposed so far are of the type shown in Figure 7.1 and fall into two main categories, according to how the multi-operand modular adder is implemented—those that use a tree of two-input modulo-M adders, realized as ROMs or in combinational logic, and those that use combinational logic, with, perhaps, a tiny amount of ROM. In the first category the outer modular reduction in Equation 7.2 is performed incrementally through the tree, since each modulo-M adder produces a result that is less than M. And in the second category that modular reduction is performed after the intermediate sum has been computed. It should be nevertheless be noted that there have been a few architectures that do not fit neatly into this broad classification.

The use of two-input modular adders realized as ROMs or combinational logic produces a structure whose performance can be rather low: the combinational adders, for example, must be full carry-propagate adders (CPA). For high performance, a better approach is to utilize a fundamental technique: in the implementation of high-performance multipliers: in a sequence of additions, carries need not be propagated with each addition but may be saved and assimilated after the last addition. Thus the multi-operand modular adder may be implemented as a tree of carry-save adders (CSAs) with a single final CPA to assimilate the partial-carry/partial-sum output of the CSA-tree. The difficulty then is how to perform the final modular reduction. The CSA+CPA approach has been used in a few designs, e.g. [1, 4, 6], some of which are described below.

7.1.1 Pseudo-SRT implementation

In the first of the two categories indicated above, Equation 7.2 is realized rather directly. A different approach to the design of the required multi-operand modular adder is to start with the basic definition of a residue; that is, as the remainder from an integer division. So we want to obtain X as

$$X = \sum_{i=1}^{N} |x_i X_i|_M - qM \qquad \text{for some integer } q$$

$$\stackrel{\triangle}{=} \widetilde{X} - qM \qquad (7.2)$$

subject to the condition

$$|\widetilde{X} - qM| < M \qquad (7.3)$$

As it stands, in order to apply Equation 7.3 directly, we have to know the value of q, and this problem is not much different from that of determining a quotient in division. There are many division algorithms that can be used to determine q, but, given that we would like to avoid division (which is a complex operation), what we seek is an algorithm that (hopefully) tells us exactly what multiple of M to subtract in Equation 7.3. Now, determining the exact value of q requires a division; so it may be that the best we can do easily is to find an approximation that is not too far off and which can easily be adjusted to yield the correct result; for example, an approximation that is off by at most unity. Such "reasonable guesswork" combined with "small corrections" are inherent in the SRT-division algorithm (Chapter 5), and the fundamental idea can be used here [4].

SRT division (Chapter 5) is an iterative process that may be partially expressed By the recurrence

$$R^{(i+1)} = rR^{(i)} - q_i D \qquad (7.4)$$

where $R^{(i)}$ is the ith partial remainder and $R^{(0)}$ is the dividend, q_i is the ith quotient digit, and r is the radix of computation (i.e. q_i is of $\log_2 r$ bits). The digit-set from which q_i is selected is a redundant-signed-digit one, $\{-a, -(a-1), \ldots 1, 0, 1, a-1, a\}$, where $r/2 \leq a \leq r-1$; and the degree of redundancy, ρ, is defined as $a/(r-1)$. The final quotient, Q, is obtained as

$$Q = \sum_{i=1}^{n/\log_2 r} q_i r^{-i}$$

although in implementation the addition is only implicit. The rationale for the use of redundant-signed-digits is that if q_i is incorrect (i.e. the subtraction in Equation 7.3 leaves a negative result), then a correction is achieved through an appropriate choice of q_{i+1}. The speed in SRT division comes from the fact that the implicit borrow-propagate subtractor (carry-propagate adder) may be replaced with a borrow-save subtractor (carry-save adder), which is much faster. Of course, this means that q_i cannot be determined precisely, but a good enough approximation can be obtained by propagating borrows (carries) in a few significant bits of the partial remainder, which will now be in partial-borrow/partial-difference (partial-carry/partial-sum) form. The result of this assimilation is then used to select one of several pre-computed multiples to subtract. Careful determination of the number of bits assimilated ensures that any incorrect "guess" is not too far off and, therefore, that easy correction is possible.

In order to ensure that $R^{(i)}$ is within proper bounds and that q_i will not be "too wrong", q_i is chosen so that

$$\left| R^{(i)} - q_i D \right| < \rho D$$

that is

$$\left|R^{(i)} - q_i D\right| < \frac{a}{r-1}D \qquad (7.5)$$

Note that Equation 7.6 has the same general form as Equation 7.4; q corresponds to q_i and M corresponds to D.

The general idea of SRT division may therefore be used as follows to realize CRT-based reverse conversion. Given the similarities indicated above, the CRT equation that corresponds to Equation 7.6 is

$$\left|\widetilde{X} - qM\right| < \frac{\tilde{a}}{r-1}M$$

where $\tilde{a}$ is determined by the largest multiple of M that may be subtracted from $\widetilde{X}$. Suppose we can ensure that the subtracted multiple of M will be either qM or $(q+1)M$. Then we may proceed by simultaneously performing both subtractions (which would require two subtractors, for speed) and then selecting the result of one of the two: $\widetilde{X} - (q+1)M$, if that is positive, and $\widetilde{X} - qM$, if the former is negative. Alternatively, some means may be used to ensure that the correct multiple is subtracted right away. We next turn to how, in essence, the value q can be determined.

In SRT division, because the divisor is variable, a quotient-digit is selected by comparing a few high-order bits of the partial remainder with a few high-order bits of the divisor. In reverse conversion, on the other hand, the "divisor" (M) is constant; so such a comparison is not necessary, but some high-order bits of the "partial remainder" ($\widetilde{X}$) must still be examined. For N moduli, number of high-order bits of $\widetilde{X}$ that must be examined is determined by r above and is

$$k + \log_2 N \qquad (7.6)$$

where k depends on the value of M.

Figure 7.2 shows an architecture for CRT reverse conversion based on the pseudo-SRT approach. The values from the ROMs are then added in a carry-save-adder tree, whose outputs are a partial-sum (PS) and a partial-carry (PC). In order to select the appropriate multiple of M, several most significant PC/PS bits are assimilated in a small carry-propagate adder (the mini-CPA) and the result then used to address a ROM that contains a pair of values corresponding to possible values of q and $q+1$, one of which is chosen. In order to know exactly which of these two is the correct one, it is necessary to know whether $\widetilde{X}$ is odd or even, and the same for X—that is, whether the multiple to be subtracted should be odd or even. This is done by examining the least significant bit of each, denoted by $p(\widetilde{X})$ and

$p(X)$. For the latter, Figure 7.2 shows this explicitly (which is not done in [4]). Similarly, in Figure 7.2, the three-input subtractor is shown explicitly for what it actually is—a combination of a CSA and a CPA.

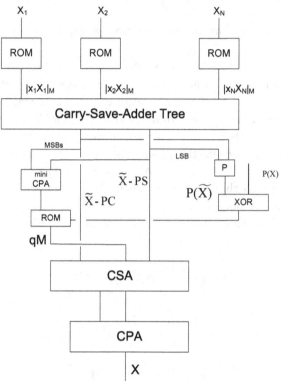

Figure 7.2: Pseudo-SRT reverse converter

7.1.2 Base-extension implementation

We now show how base-extension [2] can be used as the basis for an architecture that is cheaper (by gate count) and faster (by gate delay) than that of the pseudo-SRT approach. Observe that in Equation 7.2, each $|x_j X_j|_M$ is less than M; therefore, $\sum_{i=1}^{N} |x_j X_j|_M < NM$. And from Equation 7.3, we have

$$X = \sum_{j=1}^{N} |x_j X_j|_M - qM$$
$$< NM - qM$$

with $X < M$, according to the condition of Equation 7.4. It therefore follows that $q < N$.

Suppose we now pick an extra, *redundant*[1] modulus, m_E, such that $m_E \geq N$. Base-extension here then consists of the computation of $x_E \stackrel{\triangle}{=} |X|_{m_E}$. We will assume that x_E is readily available, even though during reverse conversion we do not know what X is, nor do we wish to have to compute x_E at that stage. This assumption is reasonable because x_E is easily computed during the pre-arithmetic forward conversion used to obtain $x_i = |X|_{m_i}$, $i = 1, 2, \ldots, N$. Indeed, in what follows, we shall arrange matters so that in contrast with the x_is, no actual computation will be required to obtain x_E.

Reducing both sides of Equation 7.3, with respect to the modulus m_E, we have

$$|X|_{m_E} = \left| \sum_{j=1}^{N} |x_j X_j|_M - qM \right|_{m_E}$$
$$\stackrel{\triangle}{=} \left| \widetilde{X} - qM \right|_{m_E}$$

whence

$$|qM|_{m_E} = \left| \widetilde{X} - x_E \right|_{m_E}$$

and multiplying through by $\left| M^{-1} \right|_{m_E}$ yields

$$|q|_{m_E} = \left| \left| M^{-1} \right|_{m_E} \left(\left| \widetilde{X} \right|_M - x_E \right) \right|_{m_E}$$

Now, $q < N$ and $m_E \geq N$, so $|q|_{m_E} = q$. Therefore the last equation becomes

$$q = \left| \left| M^{-1} \right|_{m_E} \left(\left| \widetilde{X} \right|_{m_E} - x_E \right) \right|_{m_E} \tag{7.7}$$

Once the value of q has been determined, a multiplication and a subtraction yield the sought number X, by Equation 7.3. And if all the possible values

---

[1]That is, the modulus goes not extend the dynamic range determined by $m_1, m_2, \ldots, m_N$; nor does the extra modulus have any role in the residue arithmetic.

of qM have been pre-computed and stored, then it suffices to simply read out the appropriate, correct value and perform a subtraction.

Equations 7.3 and 7.8 may be implemented as follows. In order to simplify the reductions modulo m_E, we impose the additional constraint that m_E be a power of two, say 2^n; the smallest n such that $2^n \geq N$ will suffice. Then a reduction of some operand modulo m_E consists of a simple right-shift by n bits; and this is just means "reading off" those bits. So to compute $\left|\widetilde{X}\right|_{m_E}$, we simply take the n least significant bits of $\widetilde{X}$. (A similar, simple "reading" suffices to obtain x_E from X.) Since these bits of $\widetilde{X}$ will be in partial-carry(PC)/partial-sum(PS) form, they are assimilated in a small, n-bit carry-propagate adder (mini-CPA); we shall use $\widetilde{X_L}$ to refer to the result of this assimilation. We avoid the arithmetic operations required to compute qM by storing all the possible values in a ROM that is addressed by $\widetilde{X_L}$ and x_E. Since $|M^{-1}|_{m_E}$ is a constant, no actual multiplication is necessary; instead, the value of that constant is taken into account when the pre-computed values are mapped into the ROM. The complete architecture is then as shown in Figure 7.3. As in the architecture of Figure 7.2, the nominal three-input subtractor actually consists of a CSA, to reduce three inputs to two, followed by a CPA.

A final note is in order regarding the choice of m_E: since m_E is a power of two, in order for $|M^{-1}|_{m_E}$ to exist, M must be odd. This is, of course, a trivial constraint that can be achieved simply by using only odd moduli for $m_1, m_2, \ldots, m_N$.

EXAMPLE. Take $m_1 = 5, m_2 = 7$, and $m_3 = 9$. Then $M_1 = 63, M_2 = 45, M_3 = 35, \left|M_1^{-1}\right|_{m_1} = 2, \left|M_2^{-1}\right|_{m_2} = 5, \left|M_3^{-1}\right|_{m_3} = 8$, and $M = 315$. Since there are three moduli, we take $m_E = 4 = 2^2$; that is, $n = 2$. Then $\left|M^{-1}\right|_{m_E} = 3$. Now consider the representation in this system of $X = 72$. During forward conversion, we would have obtained $72 \cong \langle 2, 2, 0\rangle$ and $|X|_{m_E} = 0$. In reverse conversion we would then have[2] $X_1 = 226, X_2 = 225$, and $X_3 = 280$. So

$$\widetilde{X} = |2 \times 126|_{315} + |2 \times 225|_{315} + |0 \times 380|_{315} = 252 + 135 = 387$$

Since $n = 2$, we compute $\left|\widetilde{X}\right|_{m_E}$ as the (assimilated values of the) two

---

[2]For convenience, $\widetilde{X}$ and $\widetilde{X_L}$ are given in assimilated form. We leave it to the reader to work out the details of the unassimilated forms.

least significant bits of the representation of $\widetilde{X}$ and obtain $\widetilde{X_L} = 3$. Thus

$$q = \left|\left|M^{-1}\right|_{m_E}\left(\left|\widetilde{X}\right|_{m_E} - x_E\right)\right|_{m_E}$$
$$= |3 \times (3-0)|_4$$
$$= 1$$

and from the ROM we then obtain $qM = 315$. A final subtraction then yields the sought value of X: $X = 387 - 315 = 72$. The reader is invited to try his or her hand with different values of X and m_E.
END EXAMPLE

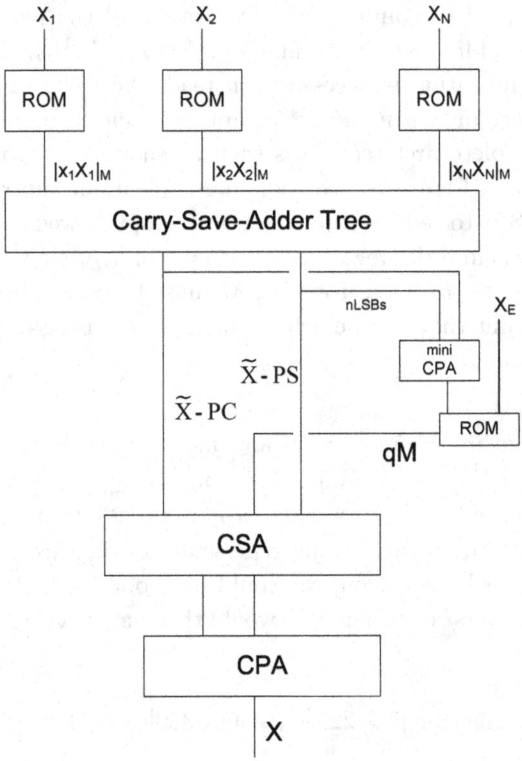

Figure 7.3: Base-extension reverse converter

The architecture of Figure 7.3 is superficially similar to that of Figure 7.2, but that is as far as it goes. The major differences are as follows. First, ROM required for the storage of the qM values is smaller in Figure 7.3. In the architecture of Figure 7.2, if there are N moduli, then the ROM will be of $2N$ entries (one entry for each possible value of $p(\widetilde{X} \oplus p(X))$. On the other hand, in Figure 7.3, exactly N entries will suffice. Second, in either architecture, the mini-CPA has an effect on the critical path, but it will be smaller in Figure 7.3—n-bits wide versus $k + \log_2 N$ (Equation 7.7). For example for 15 and 16 moduli, the mini-CPA in Figure 7.2 will be at least 8 bits wide, whereas the mini-CPA in Figure 7.3 will be 4-bit for 15 moduli and 5-bit for 16 moduli. Third, the qM-ROM in Figure 7.2 is accessed with the most significant bits (MSBs) of $\widetilde{X}$, whereas the one in Figure 7.3 is accessed with the least significant bits (LSBs). The LSBs are available earlier than the MSBs; therefore, in Figure 7.3, the operations of the mini-CPA and the qM-ROM access can be overlapped with the computation of the high-order bits of $\widetilde{X}$. Lastly, suppose we change the dynamic range by keeping the same number of moduli but using larger moduli. Then a wider mini-CPA will be required in the architecture of Figure 7.2 (because the component k in Equation 7.7 depends on M), whereas in the architecture of Figure 3 no change is required, as the adder width is determined by N only. In both architectures, the contents of all ROMs must be changed.

7.2 Mixed-radix number systems and conversion

We have already seen above that there is some relationship between representations in a residue number system and those in some mixed-radix number system. The latter are therefore significant, since they are weighted number systems and so facilitate the implementation of operations (such as magnitude comparison) that are problematic in residue number systems. In this section we shall formulate a mixed-radix numbers system and show its application in reverse conversion.

Consider the example of a number system in which the digit-weights are not fixed but vary according to the sequence $15, 13, 11$, with the interpretation that the weights of the digits are, from left to right, $15 \times 13 \times 11 \times 7, 13 \times 11 \times 7, 11 \times 7, 7$ and 1. Essentially, each of these weights determines a distinct radix, whence the "mixed-radix". To ensure unique representations in such a system, it is necessary to impose the constraint that the maximum weight contributed by the lower k digits must never exceed the positional weight of the $(k+1)$st digit. Then, if the radices are $r_N, r_{N-1}, \ldots r_1$, any number X can be uniquely expressed in mixed-radix form as [8],

$$X \cong (z_N, z_{N-1}, z_{N-2}, \ldots z_1)$$

whose interpretation is that

$$X = z_N r_{N-1} r_{N-2} \cdots r_1 + \cdots + z_3 r_2 r_1 + z_2 r_1 + z_1 \qquad (7.8)$$

where $0 \leq z_i < r_i$. It is evident that this is a weighted, positional number system. In what follows, "mixed-radix number system" will indicate this particular formulation.

The bounds above on z_i guarantee a unique representation. The conversion from a residue number to a mixed-radix number may be regarded as a reverse transform since the mixed-radix system is weighted. We next show how to obtain the digits z_i, and, therefore, how to reverse-convert to the conventional equivalent, X.

Given the above, the association between a mixed-radix representation (Equation 7.9) and a residue representation, $\langle x_N, x_{N-1}, \ldots, x_1 \rangle$, with respect to moduli $m_1, m_2, \ldots, m_N$, we make the r_is in Equation 7.9 correspond to the m_is. That equation then becomes

$$X = z_N m_{N-1} m_{N-2} \cdots m_1 + \ldots + z_3 m_2 m_1 + z_2 m_1 + z_1 \qquad (7.9)$$

A modular reduction, with respect to m_1, of both sides of this equation yields

$$|X|_{m_1} = z_1$$
$$= x_1$$

To find z_2, we re-write Equation 7.10 as

$$X - z_1 = z_N m_{N-1} m_{N-2} \cdots m_1 + \ldots + z_3 m_2 m_1 + z_2 m_1$$

and a reduction modulo m_2 then yields

$$|X - z_1|_{m_2} = |z_2 m_1|_{m_2}$$

Multiplying through by $\left|m_1^{-1}\right|_{m_2}$ yields

$$\left|\left|m_1^{-1}\right|_{m_2} (X - z_1)\right|_{m_2} = |z_2|_{m_2}$$
$$= z_2 \qquad \text{since } z_2 < m_2$$

And since

$$|X - z_1|_{m_2} = \left||X|_{m_2} - |z_1|_{m_2}\right|_{m_2}$$
$$= |x_2 - z_1|_{m_2} \qquad \text{since } z_1 < m_2$$

by Equation 2.2
$$z_2 = \left|\left|m_1^{-1}\right|_{m_2}(x_2 - z_1)\right|_{m_2}$$

We apply a similar process to obtain z_3:
$$|X - (z_2 m_1 + z_1)|_{m_3} = |z_2 m_2 m_1|_{m_3}$$

whence
$$z_3 = \left|\left|(m_2 m_1)^{-1}\right|_{m_3}(X - (z_2 m_1 + z_1))\right|_{m_3}$$

And from
$$|(X - (z_2 m_1 + z_1))|_{m_3} = \left||X|_{m_3} - |+z_1|_{m_3}\right|_{m_3}$$
$$= |x_3 - (z_2 m_1 + z_1)|_{m_3}$$

we have
$$z_3 = \left|\left|(m_2 m_1)^{-1}\right|_{m_3}(x_3 - (z_2 m_1 + z_1))\right|_{m_3}$$

Continuing this process, the mixed-radix digits, z_i, can be retrieved from the residues as

$$z_1 = x_1 \qquad (7.10)$$

$$z_2 = \left|\left|m_1^{-1}\right|_{m_2}(x_2 - z_1)\right|_{m_2}$$

$$z_3 = \left|\left|(m_1 m_2)^{-1}\right|_{m_3}(x_3 - (z_2 m_1 + z_1))\right|_{m_3}$$

$$\vdots$$

$$z_N = \left|(m_1 m_2 \ldots m_{N-1})^{-1}\right|_{m_N} |(x_n - (x_{N-1} m_{N-2} \ldots z_2 m_1 + z_1))|_{m_N}$$

The N multiplicative inverses required in Equations 7.11 are constants and so can be pre-computed and stored beforehand.

This last set of equations shows that the mixed-radix approach is inherently a sequential approach: it is necessary to determine z_1 first, then z_2, then z_3, and so on. On the other hand, with the CRT approach, as outlined above, the partial sums X_js (which roughly correspond to the z_js here) can be computed in parallel. The use of the CRT requires a final modular reduction, which can be done only at the end, when the all X_js have been computed; but this is the only serial aspect, and it is a relatively minor one.

Reformulations of Equations 7.11 can allow the exploitation of some more parallelism. One such reformulation is as follows. From an application of Equation 2.2, we have

$$\left|\left|m_i^{-1}\right|_{m_k}\left|m_j^{-1}\right|_{m_k}\right|_{m_k} = \left|(m_i m_j)^{-1}\right|_{m_k}$$

We may therefore rewrite the equations above for z_3 to z_N—those for z_2 and z_1 are unchanged—into

$$z_3 = \left|\left|\left|m_2^{-1}\right|_{m_3}\left|m_1^{-1}\right|_{m_3}\right|_{m_3}(x_3 - (z_2 m_1 + z_1))\right|_{m_3} \quad (7.11)$$

$$= \left|\left|m_2^{-1}\right|_{m_3}\left(\left|m_1^{-1}\right|_{m_3}(x_3 - z_1) - z_2\right)\right|_{m_3}$$

$$\vdots$$

$$z_N = \left|\left|m_{N-1}^{-1}\right|_{m_N}\left(\left|m_{N-2}^{-1}\right|_{m_N}\left(\cdots\left|m_2^{-1}\right|_{m_N}\left(\left|m_1^{-1}\right|_{m_N}(x_N - z_1) - z_2\right)\right.\right.\right.$$
$$\left.\left.\left.\cdots\right) - z_{N-1}\right)\right|_{m_N}$$

Consider now the computation of z_3, for example. With Equations 7.11 all sub-computations must be carried out in sequence (ignoring the obvious slight parallelism that comes from distribution of multiplication over addition). On the other hand, with Equations 7.12, the computation of $\left|m_1^{-1}\right|_{m_3}(x_3 - z_1)$ can be started as soon as z_1 is available. Other than the parallelism, the main difference between Equations 7.11 and 7.12 is in the number of multiplicative inverses required —N for the former and $N(N-1)/2$ for the latter. An example is given next. In this and other similar examples, negative numbers are given as appropriate additive inverses (see Section 2.3.1).

EXAMPLE. Suppose we wish to find the number, X, whose residue representation is $\langle 1, 0, 4, 0\rangle$ relative to the moduli-set $\{2, 3, 5, 7\}$. Applying Equations 7.12, we get

$$z_1 = 1$$

$$z_2 = \left||2^{-1}|_3 (0-1)\right|_3$$
$$= |2 \times -1|_3$$
$$= |2 \times 2|_3 \quad \text{additive inverse of } -1 \text{ w.r.t 3 is 2}$$
$$= 1$$

$$z_3 = \left||3^{-1}|_5 \left(|2^{-1}|_5 (4-1) - 1\right)\right|_3$$
$$= |2 \times (3 \times 3 - 1)|_5$$
$$= 1$$

$$z_4 = \left||5^{-1}|_7 \left(|3^{-1}|_7 \left(|2^{-1}|_7 (0-1) - 1\right) - 1\right)\right|_7$$
$$= |27|_7$$
$$= 6$$

Therefore

$$X \cong (1,1,1,6)$$

and for the conventional form, we translate this as

$$X = 6 \times 2 \times 3 \times 5 + 1 \times 2 \times 3 + 1 \times 2 + 1$$
$$= 189$$

END EXAMPLE

An architecture for the implementation of Equations 7.12 is shown in Figure 7.4. Here, we assume the use of sum-addressable ROMs [15] to produce the product of differences and multiplicative inverses and the use of ordinary ROMs to produce of the products of moduli and the zs. A tree of carry-save adders and one carry-propagate adder computes the summation of Equation 7.10. We leave it to the reader to formulate a design of only combinational logic.

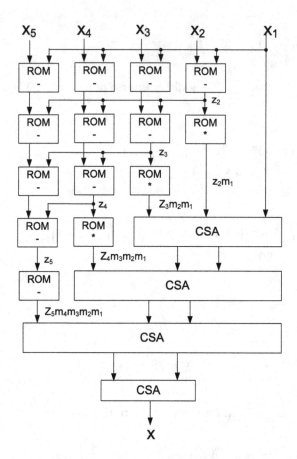

Figure 7.4: Mixed-radix reverse converter ($N = 5$)

If, as is most likely to be the case, the reverse conversion is required for the results of RNS arithmetic units, then it is possible and convenient to carry out the computations of Equation 7.12 in such units. That is, the equations can be formulated to naturally suit arithmetic units that are capable of multiplications of the form $\langle x_1, x_2, \ldots, x_N \rangle \times \langle y_1, x_y, \ldots, y_N \rangle$ and additions of the form $\langle x_1, x_2, \ldots, x_N \rangle + \langle y_1, x_y, \ldots, y_N \rangle$. For this, we proceed as follows.

Let U be the residue number for which we seek a mixed-radix representation $(z_1, z_2, \ldots, z_N)$. And define the residue numbers Y_i by

$$Y_1 = U$$
$$Y_{i+1} = \left\langle \left|m_i^{-1}\right|_{m_1:m_N} \right\rangle (Y_i - Z_i) \qquad (7.12)$$
$$z_i = |y_i|_{m_i}$$

where Z_i is the residue number $\langle z_i, z_i, \ldots, z_i \rangle$, $\left\langle \left|m_i^{-1}\right|_{m_1:m_N} \right\rangle$ is the residue number $\langle \left|m_i^{-1}\right|_{m_1}, \left|m_i^{-1}\right|_{m_2}, \ldots, \left|m_i^{-1}\right|_{m_N} \rangle$, and y_i is digit i of Y_i. Once z_i has been determined the corresponding residue digits in the other residue numbers may be discarded; we shall use * to indicate this in the following example. As in the preceding example, appropriate additive inverses are used for negative numbers.

EXAMPLE. Suppose we wish to find the number, X, whose representation is $\langle 1, 0, 4, 0 \rangle$ relative to the moduli-set $\{2, 3, 5, 7\}$. Applying Equations 7.13, we get

$$Y_1 = \langle 1, 0, 4, 0 \rangle$$
$$z_1 = 1$$
$$\left\langle \left|m_1^{-1}\right|_{m_1:m_4} \right\rangle = \langle *, 2, 3, 4 \rangle$$

$$Y_2 = \langle *, 2, 3, 4 \rangle \times \langle *, -1, 3, -1 \rangle$$
$$= \langle *, 2, 3, 4 \rangle \times \langle *, 2, 3, 6 \rangle$$
$$= \langle *, 1, 4, 3 \rangle$$
$$z_2 = 1$$
$$\left\langle \left|m_2^{-1}\right|_{m_1:m_4} \right\rangle = \langle *, *, 2, 5 \rangle$$
$$Y_3 = \langle *, *, 2, 5 \rangle \times \langle *, *, 3, 2 \rangle$$
$$= \langle *, *, 1, 4 \rangle$$

$$z_3 = 1$$
$$\left\langle \left|m_3^{-1}\right|_{m_1:m_4} \right\rangle = \langle *, *, *, 3 \rangle$$

$$Y_4 = \langle *, *, *, 3 \rangle \times \langle *, *, *, 2 \rangle$$
$$= \langle *, *, *, 6 \rangle$$
$$z_4 = 6$$

whence $X \cong (1,1,1,6)$, and $X = 189$, as in the last example.
END EXAMPLE

Because the mixed-radix system is a weighted, positional number system, and is, in that sense, similar to conventional number systems, operations such as magnitude-comparison are much easier than in residue number systems. For example, given the mixed-radix numbers $X \cong (x_N, x_{N-1}, \ldots, x_1)$ and $Y \cong (y_N, y_{N-1}, \ldots, y_1)$, we may immediately deduce that $X > Y$ if $x_N > y_N$.

Overflow-detection, which in fact is equivalent to magnitude-comparison, is another of the more problematic operations in residue number systems. Mixed-radix conversion may be used to facilitate the implementation of this operation, as follows. Choose a redundant, extra modulus, m_E, and base-extend the given residue representation of a number, X, by computing $x_E \triangleq |X|_{m_E}$. Then overflow has occurred if $x_E \neq 0$. To compute x_E, Equation 7.9 is replaced with

$$X = z_E m_N m_{N-1} \cdots m_1 + z_N m_{N-1} m_{N-2} \cdots m_1 + \cdots + x_2 m_1 + z_1$$

and the z_is are computed as above. Then x_E is obtained as

$$x_E = |z_E m_N m_{N-1} \cdots m_1 + z_N m_{N-1} m_{N-2} \cdots m_1 + \cdots + x_2 m_1 + z_1|_{m_E}$$

7.3 The Core Function

The Core Function provides a method for reverse conversion that is reasonably straightforward, although there are some minor complications [6]. Recall, from Chapter 2, that the core, $C(X)$, of a number X represented by the residues $\langle x_1, x_2, \ldots, x_n \rangle$, relative to the moduli $m_1, m_2, \ldots, m_N$, i.e. $x_i = |X|_{m_i}$, is defined as

$$C(X) = \sum_{i=1}^{N} \frac{w_i}{m_i} X - \sum_{i=1}^{N} \frac{w_i}{m_i} x_i \qquad \text{for certain weights, } w_i$$

$$= \frac{C(M)}{M} X - \sum_{i=1}^{N} \frac{w_i}{m_i} x_i \qquad \text{where } M = \prod_{i=1}^{N} m_i \qquad (7.13)$$

from which we have

$$X = \frac{M}{C(M)} C(X) + \sum_{i=1}^{N} M_i w_i \frac{M_i}{C(M) x_i} \qquad \text{where } M_i = M/m_i \quad (7.14)$$

This, therefore, provides a method for obtaining X from its residues, which method does not have the drawbacks of the Chinese Remainder Theorem or the Mixed-Radix Conversion. But the application of such a method requires two things. The first is that there be an efficient method to compute cores; the other other is some way to deal with the potentially difficult operation of (the implied) divisions. The division can be eliminated by either choosing $C(M)$ to be a power of two, in which case all that is required is right-shifting, or by choosing $C(M)$ to be one of the moduli. In the latter case, if $C(M) = m_j$, then

$$X = M_j C(X) + w_j \frac{M_j}{m_j} x_j + \sum_{i=1, i \neq j}^{N} w_i \frac{M_i}{m_j} x_j$$

And if ensure that $w_j = 0$, then all division operations are eliminated:

$$X = M_j C(X) + \sum_{i=1, i \neq j}^{N} w_i \frac{M_i}{m_j} x_j$$

Because each M_i is divisible by m_j if $i \neq j$.

To compute the core, a CRT for Core Functions is employed. Recall that the CRT relates a number X to its residues $x_1, x_2, \ldots, x_N$, with respect to the moduli $m_1, m_2, \ldots, m_N$, as follows

$$X = \left| \sum_{i=1}^{N} x_i X_i \right|_M \qquad \text{where } X_i = M_i \left| M_i^{-1} \right|_{m_i}$$

$$= \sum_{i=1}^{N} x_i X_i - qM \qquad \text{for some integer } q$$

Substituting for X in the definition of the Core Function (Equation 7.14)

$$C(X) = \sum_{i=1}^{N} \frac{w_i}{m_i} \left(\sum_{j=1}^{N} x_j X_j - qM \right) - \sum_{i=1}^{N} \frac{w_i}{m_i} x_i$$

$$= \sum_{i=1}^{N} \frac{w_i}{m_i} \sum_{j=1}^{N} x_j X_j - \sum_{i=1}^{N} \frac{w_i}{m_i} x_i - \sum_{i=1}^{N} q \frac{w_i}{m_i} M$$

$$= \sum_{i=1}^{N} \frac{w_i}{m_i} \sum_{j=1}^{N} x_j X_j - \sum_{i=1}^{N} \frac{w_i}{m_i} x_i - qC(M) \qquad \text{by Equation 2.13}$$

$$= \sum_{j=1}^{N} x_j X_j \sum_{i=1}^{N} \frac{w_i}{m_i} - \sum_{j=1}^{N} \frac{w_j}{m_j} x_j - \sum_{i=1}^{N} qC(M) \qquad (7.15)$$

Now the core, $C(X_j)$, of $M_j \left|M_j^{-1}\right|_{m_j}$ is, by definition,

$$C(X_j) = \sum_{i=1}^{N} \frac{w_i}{m_i} X_j - \left|\sum_{i=1}^{N} \frac{w_i}{m_i} X_j\right|_{m_i}$$

Since

$$|X_j|_{m_i} \triangleq \left|M_j \left|M_j^{-1}\right|_{m_j}\right|_{m_i} = \begin{cases} 1 \text{ if } i = j \\ 0 \text{ otherwise} \end{cases}$$

this core is

$$C(X_j) = X_j \sum_{i=1}^{N} \frac{w_i}{m_i} - \frac{w_j}{m_j}$$

whence

$$\sum_{j=1}^{N} x_j C(X_j) = \sum_{j=1}^{N} x_j X_j \sum_{i=1}^{N} \frac{w_i}{m_i} - \sum_{j=1}^{N} x_j \frac{w_j}{m_j}$$

and, finally, substitution into Equation 7.16 yields

$$C(X) = \sum_{j=1}^{N} x_j C(X_j) - qC(M) \tag{7.16}$$

which is the CRT for Core Functions. Since q depends on X, this equation is not, of itself, particularly useful for the computation of cores. But if both sides are reduced modulo $C(M)$, then that dependence is removed:

$$|C(X)|_{C(M)} = \left|\sum_{j=1}^{N} x_j C(X_j)\right|_{C(M)}$$

The cores $C(X_j)$ are constants that may be pre-computed and stored beforehand.

There are cases in which Equation 7.17 can be used to compute cores without any complications. But in general, this is not so. Three possible cases can arise in the application of the equation:

- $C(X)$ is within the correct range;
- $C(X) < 0$, in which case $|C(X)|_{C(M)} = C(X) + C(M)$;
- $C(X) > C(M)$, in which case $|C(X)|_{C(M)} = C(X) - C(M)$;

That is, the value obtained may be an incorrect, out-of-range value, or a correct, in-range value. Determining which of the two is the case is not an easy task. This problem may be dealt with as follows.

Chapter 2 describes the use of a "parity bit", which involves the computation of $|C(X)|_2$. And from this, we can obtain $|C(X)|_{C(M)}$ by adding or subtracting $C(M)$. But, deciding which to do is a difficult operation. To get around this, we may proceed as follows. Compute the residues, of $\lfloor X/2 \rfloor$, i.e. $|(x_i - p)/2|_{m_i}$, where p is the parity of X. Then

$$\left| C\left(\left\lfloor \frac{X}{2} \right\rfloor \right) \right|_{C(M)} = \left| \sum_{i=1}^{N} \left| \frac{x_i - p}{2} \right|_{m_i} C(X_i) \right|_{C(M)}$$

whence

$$C(X) = 2C\left(\left\lfloor \frac{X}{2} \right\rfloor \right) + \left| \sum_{i=1}^{N} w_i \left(|x_i|_2 \oplus p \right) \right|_{C(M)}$$

Architectures for the implementation of Equation 7.17 are described in [4]. Equation 7.17 shows the main merits of the Core-Function approach: it does not have the sequential aspects of MRC; nor does it require the large modular reductions of the CRT. On that basis, it appears that it ought to be easy to devise Core-Function architectures that are superior (with respect to both cost and performance) to those for the MRC and the CRT. Unfortunately, closer examination of that equation shows that not to be necessarily so. The hardware costs, for example, are likely to be rather high. and this is an area where further research is required.

7.4 Reverse converters for $\{2n-1, 2n, 2n+1\}$ moduli-sets

Reverse converters for the $\{2n - 1, 2n, 2n + 1\}$ moduli-sets, including the special moduli $2^m - 1, 2^m$, and $2^m + 1$, can be easily derived from the CRT; but here simple closed-form solutions for multiplicative inverses are possible, and these are to be preferred [7]. These inverses can subsequently be used to reduce the complexity of reverse conversions. In a $\{2n-1, 2n, 2n+1\}$ moduli-set, the weights in the equation $X = \sum_i w_i x_i$ (Section 3.1) can be determined with little computational complexity. The weights corresponding to the three moduli $m_3 \stackrel{\triangle}{=} 2n - 1, m_2 \stackrel{\triangle}{=} 2n$, and $m_1 \stackrel{\triangle}{=} 2n + 1$ may be computed as

$$w_1 = \frac{(m_1+1)m_2 m_3}{2}$$

$$w_2 = m_1(m_2-1)m_3$$

$$w_3 = \frac{m_1 m_2(m_3+1)}{2}$$

We next show shown that w_1, w_2, w_3 are indeed the correct weights.

Since each of $w_1, w_2,$ and w_3 involves m_2, m_3, $m_1 m_3$, and $m_1 m_2$ respectively, we need only show that each of $|w_1|_{m_1}$, $|w_2|_{m_2}$ and $|w_3|_{m_3}$ is 1:

$$w_1 = \frac{(2n+2) \times 2n \times (2n-1)}{2}$$

$$|w_1|_{m_1} = \left|\frac{(2n+1+1)(2n+1-1)(2n+1-2)}{2}\right|_{2n+1}$$

$$= \left|\frac{|(2n+1+1)|_{2n+1}|(2n+1-1)|_{2n+1}|(2n+1-2)|_{2n+1}}{2}\right|_{2n+1}$$

$$= \left|\frac{(1)(-1)(-2)}{2}\right|_{2n+1}$$

$$= 1$$

which proves that w_1 is the multiplicative inverse of m_1. Similarly,

$$w_2 = (2n+1) \times (2n-1) \times (2n-1)$$

$$|w_2|_{m_2} = |(2n+1)(2n-1)(2n-1)|_{2n}$$

$$= ||(2n+1)|_{2n}|(2n-1)|_{2n}|(2n-1)|_{2n}|_{2n}$$

$$= |(1)(-1)(-1)|_{2n}$$

$$= 1$$

Lastly,

$$w_3 = \frac{(2n+1) \times 2n \times (2n)}{2}$$

$$|w_3|_{m_3} = \left| \frac{(2n+1)(2n)(2n)}{2} \right|_{2n-1}$$

$$= \left| \frac{|(2n-1+2)|_{2n-1}|(2n-1+1)|_{2n-1}|(2n-1+1)|_{2n-1}}{2} \right|_{2n-1}$$

$$= \left| \frac{(2)(1)(1)}{2} \right|_{2n-1}$$

$$= 1$$

Once the weights have been computed, the conventional equivalent, X, of the residue number $\langle x_1, x_2, x_3 \rangle$, where $x_i = |X|_{m_i}$, is obtained as

$$X = \left| \sum_{i=1}^{3} w_i x_i \right|_{m_1 m_2 m_3} \quad (7.17)$$

The conversion procedure just described can implemented directly by computing the weights as $w_1 = (M + m_2 m_3)/2, w_2 = M - m_1 m_3$, and $w_3 = (M + m_1 m_2)$, which permits sharing of the common value $M \triangleq m_1 m_2 m_3$. (Note that the division by 2 is just a one-bit right shift). But it is evident that, because of the term M, such an implementation will require large multipliers to finally obtain X. A few modifications are possible that simplify the multipliers. These modifications, which we next describe, also eliminate the final modular reduction.

Take the conventional equivalent X, given by Equation 7.18. Substituting for the weights, we have,

$$|X|_M = |A + m_1(m_2 - 1)m_2 x_2 + C|_M \quad (7.18)$$

where

$$A = \frac{(m_1 + 1)m_2 m_3}{2} x_1$$

$$C = \frac{m_1 m_2 (m_3 + 1)}{2} x_3$$

Substituting M for $m_1 m_2 m_3$ and simplifying Equation 7.19, we have

$$|X|_M = \left|\widehat{A} + (M - m_1 m_3) x_2 + \widehat{C}\right|_M$$
$$= \left|\widehat{A} - m_1 m_3 x_2 + \widehat{C}\right|_M$$

where

$$\widehat{A} = \left(\frac{M}{2} + \frac{m_2 m_3}{2}\right) x_1$$

$$\widehat{C} = \left(\frac{M}{2} + \frac{m_1 m_2}{2}\right) x_3$$

Equation 7.19 can be simplified further to

$$|X|_M = \left|\frac{M}{2}(x_1 + x_3) + \frac{m_2 m_3}{2} x_1 - m_1 m_3 x_2 + \frac{m_1 m_2}{2} x_3\right|_M \quad (7.19)$$

The first term, $x_1 + x_3$, in this equation can be either even or odd. If it is even then

$$\left|\frac{M}{2}(x_1 + x_3)\right|_M = 0$$

and it is odd, then

$$\left|\frac{M}{2}(x_1 + x_3)\right|_M = \frac{M}{2}$$

So if $(x_1 + x_3)$ is even, then Equation 7.20 may be rewritten into

$$|X|_M = \left|\frac{m_2 m_3}{2} x_1 - m_1 m_3 x_2 + \frac{m_1 m_2}{2} x_3\right|_M$$

$$\triangleq |M_1 x_1 M_2 x_2 + M_3 x_3|_M \quad (7.20)$$

and if $(x_1 + x_3)$ is odd, then we have

$$|X|_M = \left|\frac{M}{2} + \frac{m_2 m_3}{2} x_1 - m_1 m_3 x_2 + \frac{m_1 m_2}{2} x_3\right|_M$$

$$\triangleq \left|\frac{M}{2} + M_1 x_1 M_2 x_2 + M_3 x_3\right|_M \quad (7.21)$$

Observe that the numbers involved in the multiplications of Equations 7.20 and 7.21 are smaller than those in the original equations. Therefore, in an implementation, the corresponding multipliers will be accordingly smaller.

A basic architecture for the implementation of Equations 7.20 and 7.21 above is shown in Figure 7.5. The value $x_1 + x_3$ need not actually be computed, since all that is required is the knowledge of whether it is odd or even; therefore, a half-adder on the least significant bits is sufficient. The sum-output bit of this half-adder is then used to select either 0 or $M/2$, according to whether that bit is 0 or 1. The generic architecture of Figure 7.5 may be refined in various ways, according to the different adder and multiplier designs of Chapters 4 and 5. We consider just one such refinement and leave it to the reader to work out others.

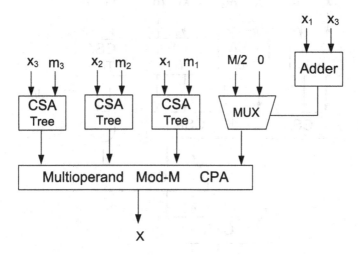

Figure 7.5: Generic reverse converter for $\{2n-1, 2n, 2n+1\}$ moduli-set

Suppose we wish to design a high-performance variant of Figure 7.5. Each multiplier may be realized as a Wallace carry-save-adder (CSA) tree, with the outputs left in unassimilated partial-carry/partial-sum. Two levels of CSAs are then used to reduce the six unassimilated outputs to three. An examination of Equations 7.21 and 7.22 shows that either a subtraction or an addition of M may be required to obtain the final result. So the nominal four-input adder of Figure 7.5 may be implemented as follows. Two CSAs are used to reduce the three "multiplier" outputs and the "mux' output to two. A modulo-M adder, which will be a carry-propagate adder (CPA), then assimilates the latter two outputs and produces the final result. The modulo-M adder nominally consists of three CPAs and a multiplexer. The

CPAs, given inputs A and B, compute $A+B$, $A+B-M$, and $A+B+M$. The MUX then selects which of the three is the correct one: if $A+B$ is negative, then $A+B+M$ is selected; if $A+B$ is positive and $A+B-M$ is negative, then $A+B$ is selected; otherwise, $A+B-M$ is selected. (Chapter 4 shows that much of the logic required for the three CPAs can be shared.) The "mux" of Figure 7.5 is obviously no more than a set of AND gates. The final design is then as shown in Figure 7.6.

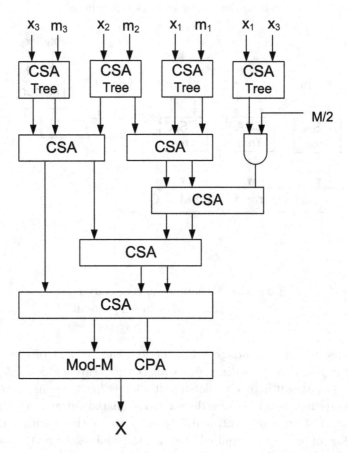

Figure 7.6: High-performance reverse converter for $\{2n-1, 2n, 2n+1\}$ moduli-set

The multiplicative inverses obtained for the $\{2n-1, 2n, 2n+1\}$ moduli-sets may be used as multiplicative inverses for the $\{2^m - 1, 2^m, 2^m + 1\}$ moduli-sets whenever n in the original set is even. For such moduli-sets, the special properties of the moduli may be exploited to eliminate the need for multipliers, and the approach described next will yield results similar to what would be obtained if w_1, w_2, and w_3 are taken as above and appropriate simplifications then carried out [10, 13]. The values required for the moduli-set $\{2^m - 1, 2^m, 2^m + 1\}$ are

$$m_3 = 2^m - 1 \qquad w_3 = 2^m(2^m + 1)2^{m-1}$$
$$m_2 = 2^m \qquad w_2 = (2^m - 1)(2^m + 1)(2^m - 1)$$
$$m_1 = 2^m + 1 \qquad w_1 = 2^m(2^m - 1)(2^{m-1} + 1)$$
$$M = (2^m - 1)2^m(2^m + 1)$$

whence, by Equation 7.1

$$X = \big|2^m(2^m + 1)2^{m-1}x_3 + (2^m - 1)(2^m + 1)(2^m - 1)x_2 + 2^m(2^m - 1)$$
$$(2^{m-1} + 1)x_1\big|_M$$

and, by Equation 7.3,

$$X + q2^m(2^{2m} - 1) = 2^m(2^m + 1)2^{m-1}x_3 + (2^m - 1)(2^m + 1)(2^m - 1)x_2$$
$$+ 2^m(2^m - 1)(2^{m-1} + 1)x_1$$

Applying integer division by 2^m and certain algebraic manipulations to this equation yields

$$\left\lfloor \frac{X}{2^m} \right\rfloor + q(2^{2m} - 1) = 2^{2m-1}x_3 + 2^{m-1}x_3 + 2^{2m}x_2 - 2^m x_2 - x_2$$
$$+ 2^{2m-1}x_1 + 2^{m-1}x_1 - x_1$$

whence

$$\left\| \left\lfloor \frac{X}{2^m} \right\rfloor \right\|_{2^{2m}-1} = \big|2^{2m-1}x_3 + 2^{m-1}x_3 + 2^{2m}x_2 \qquad (7.22)$$
$$- 2^m x_2 - x_2 + 2^{2m-1}x_1 + 2^{m-1}x_1 - x_1\big|_{2^{2m}-1}$$
$$\triangleq |Y|_{2^p-1}$$

Note that $\lfloor X/2^m \rfloor$ is just the most significant $2m$ bits of X. Also, as $\lfloor X/2^m \rfloor < 2^{2m} - 1$, we have $||\lfloor X/2^m \rfloor||_{2^p-1} = \lfloor X/2^m \rfloor$. Once these bits have been obtained, a concatenation to the least significant m bits of X yields the desired result; and the latter bits are, of course, just x_2.

For the implementation, some-bit manipulation is necessary to make the conversion-hardware efficient. Equation 7.22 can be rewritten into

$$\left\lfloor \frac{X}{2^m} \right\rfloor = |A_3 + A_2 + A_1 - x_1|_{2^p-1} \qquad (7.23)$$

where

$$A_3 = \left|\left(2^{2m-1} + 2^{m-1}\right)x_3\right|_{2^p-1}$$
$$A_2 = \left|\left(2^{2m} - 2^m - 1\right)x_2\right|_{2^p-1}$$
$$A_1 = \left|\left(2^{2m-1} + 2^{m-1}\right)x_1\right|_{2^p-1}$$

The constants A_3, A_2 and A_1 can be evaluated the property that $\left|I \times 2^l\right|_{2^p-1}$ is equivalent to circular shifting the binary representation of the integer I expressed in p bits, l positions to the left. As an example, consider $\left|7 \times 2^3\right|_{15}$. This is equivalent to 0111 shifted left circularly by 3 bits as 1011. (For more details on this, see Chapter 5, on multiplication modulo $2^n - 1$.)

The constant A_2 can be simplified as

$$A_2 = \left|(2^p - 1) - |2^m x_2|_{2^p-1}\right|_{2^p-1}$$

where $p = 2m$. Expressing x_3, x_2 and x_1 as p bit binary numbers and using the above property, the constants can be evaluated with little complexity as

$$A_3 = \underbrace{b_{30}b_{3(m-1)}\ldots b_{32}b_{31}}_{m}\underbrace{b_{30}b_{3(m-1)}\ldots b_{32}b_{31}}_{m} \qquad (7.24)$$

$$A_2 = \underbrace{\bar{b}_{2(m-1)}\bar{b}_{2(m-2)}\ldots \bar{b}_{22}\bar{b}_{21}\bar{b}_{20}}_{m}\underbrace{111\ldots 111}_{m}$$

$$A_1 = \underbrace{b_x b_{1(m-1)}\ldots b_{12}b_{11}}_{m}\underbrace{b_x b_{1(m-1)}\ldots b_{12}b_{11}}_{m}$$

where $b_x = b_{10} \oplus b_{1m}$. Further, $|-x_1|_{2^p-1}$ is

$$|-x_1|_{2^p-1} = \underbrace{111\ldots 111}_{m-1}\underbrace{\bar{b}_{1,m}\ \bar{b}_{1,(m-1)}\ldots \bar{b}_{1,1}\bar{b}_{1,0}}_{m+1} \qquad (7.25)$$

In the hardware implementation of the reverse converter, evaluation of the three constants can be accomplished by reforming the bits of the three residues. The hardware for this converter requires an adder to add A_3 and A_2. The carry bit generated here has a binary weight of 2^p and upon modulo

$2^p - 1$ operation on it generates a carry bit C_0. The second adder evaluates $A_1 - x_1$ and this result will be less than $2^p - 1$. These two additions can be done in parallel. The results of the two adders are summed with C_0 in the third adder. Based on the output of the third adder, a fourth adder is necessary to sum the outputs of the third adder and the overflow carry of the third adder. Such a circuit uses four $2m$-bit carry-propagate adders.

Instead of using four $2m$-bit carry-propagate adders in three stages and an input stage for sensing the overflow, the sum in Equation 7.23 may instead be computed more quickly by using two $2m$-bit carry-save adders (in series), with end-around carries for modulo-$(2^n + 1)$ addition, and an assimilating $2m$-bit modulo-$(2^m - 1)$ carry-propagate adder [1, 12]. In the design of [1], for speed, the nominal assimilating carry-propagate adder is implemented as two parallel carry-propagate adders, one for an end-around-carry (carry-in) of 0, and the other for an end-around-carry (carry-in) of 1. A multiplexer then selects the correct input. The reverser converter therefore has a cost of $6m+1$ FAs, $2m-1$ OR gates and a $2n$ bit-multiplexer and a delay of $2t_{FA}+t_{CPA2m}+t_{MUX}$ where t_{FA} is the FA delay, t_{CPA} is the CPA delay, and t_{MUX} is the multiplexer delay. Nevertheless, it should be noted that (Section 4.3) that the assimilating carry-propagate adder can be implemented as a single parallel-prefix adder, with no MUX, thus leading to improvements in both cost and performance.

A side-effect of the approach just described is that it permits a redundant representation of 0, which, unless action is taken, is normal in modulo $2^p - 1$ arithmetic. This leads to incorrect results when the residues are of the form $< k, k, k >$. For these cases, the converter yields the decoded value as $X = 1111\ldots11k$ where k represents the m bits of x_2. The uppermost $2m$ bits are 1s instead of 0s. This can be corrected with little penalty when a parallel-prefix adder is cost.

It is possible to further reduce the number of FA and the delay by considering the constants in Equation 7.24 and the residue representation in Equation 7.25. The key observation that leads to a reduction in FA is that three of the four summands have identical lower and upper $m - 1$ fields. This is easily seen in constants A_3 and A_1. However, in the case of A_2 and $-x_1$, the lower order m bits of A_2 and the higher order $m - 1$ bits of $-x_1$ are 1s. Reordering A_3, A_2, A_1 and $-x_1$ the similarities become fairly obvious.

$$A_3 = \underbrace{b_{30}b_{3(m-1)}\ldots b_{32}b_{31}}_{A_3(2m-1\,:\,m)} \underbrace{b_{30}b_{3(m-1)}\ldots b_{32}b_{31}}_{A_3(m-1\,:\,0)} \qquad (7.26)$$

$$\bar{A}_2 = \underbrace{111\ldots\ldots\ldots 1\bar{x}_{1,m}}_{-x_1(2m-1\,:\,m)} \underbrace{111\ldots\ldots\ldots 111}_{A_2(m-1\,:\,0)}$$

$$A_1 = \underbrace{b_x b_{1(m-1)}\ldots b_{12}b_{11}}_{A_1(2m-1\,:\,m)} \underbrace{b_x b_{1(m-1)}\ldots b_{12}b_{11}}_{A_1(m-1\,:\,0)}$$

$$-\bar{x}_1 = \underbrace{\bar{b}_{2(m-1)}\ldots \bar{b}_{22}\bar{b}_{21}\bar{b}_{20}}_{A_2(2m-1\,:\,m)} \underbrace{\bar{b}_{1,(m-1)}\ldots \bar{b}_{1,1}\bar{b}_{1,0}}_{-x_1(m-1\,:\,0)}$$

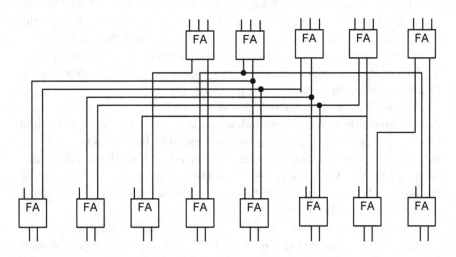

Figure 7.7: CSAs for $\{2^m - 1, 2^m, 2^m + 1\}$ reverse converter

The savings in the number of FA is obtained as follows. Once the result of addition of $A_3(m-1:1)$ of A_3, $A_2(m-1:1)$ of $\bar{A}_2$ and $A_1(m-1:1)$ of A_1 are obtained, the same result can be used for the additions of $A_3(2m-1:m+1)$ of A_3, $-x_1(2m-1:m+1)$ of $\bar{A}_2$ and $A_1(2m-1:m+1)$ of A_1. Therefore, by exploiting the symmetry in the upper and lower m bit fields, a saving of only $m-1$ adders can be achieved. This is so because, only $m-1$

and not m upper most bits of $-x_1$ are 1s. Further, the $m-1$ adders have ones as constant input as in the case of MOMA converter and therefore, these adders can be treated as OR gates. Hence this modification results in a saving of $m-1$ OR gates. The carry save adder structure is therefore as shown in Figure 7.7, for $m=4$.

The reverse converter outlined above, from [1], will have a carry-propagation of $4m$ bits. However, the carry-propagation can be limited to $2m$ bits by utilizing a $2m$-bit wide carry propagation adder with a multiplexer. This is done as follows. The final stage of the reverse converter has two $2m$-bit carry-propagate adders in parallel. Consider each of these $2m$-bit adders as a cascade of two m bit-adders, so that we nominally have four carry-propagate adders: CPA0, CPA1, CPA2 and CPA3, where CPA0 and CPA1 form the upper $2m$-bit adder, and CPA2 and CPA3 form the lower $2m$-bit adder. Since they work in parallel CPA0 and CPA2 will have identical m bit inputs, namely the high-order partial-sum bits, $\text{PC}_{2m-1:m}$, and the high-order partial-carry bits $\text{PC}_{2m-1:m}$. Similarly, CPA1 and CPA3 will be adding the low-order partial-sum and partial-carry bits. The partial-carry bits into CPA0 and CPA2 will be different, and that is also the case for CPA1 and CPA3. So, we may choose to set the carry into CPA0 to 0 and that into CPA2 to 1. A similar arrangement is done with regard to CPA1 and CPA3 carry inputs. All that is left is to take the carries out of the four m bit-adders and multiplex the correct result out. The multiplexing functions are

$$f = c_0 + c_2 c_3$$
$$g = c_1 + c_2 c_3$$

where c_i is the carry from m bit carry-propagate adder i. The total operational delay for this converter is therefore $2t_{FA} + t_{CPA(m)} + 2t_{NAND} + t_{MUX}$. The carry-propagate part of the reverse converter is shown in Figure 7.8. The final result of the conversion is obtained when the result of the addition in Equation 7.23 is shifted m bits to the left and the residue x_2 concatenated with it. Finally, as in the preceding design, that of Figure 7.8 can be beneficially improved through the use of parallel-prefix adders instead of generic adders, as has been assumed in both cases.

Another design for a fast reverse converter for the $\{2^m-1, 2^m, 2^m+1\}$ moduli-set will be found in [13]. However, in that design, the speed is obtained at the cost of rendering unusable part of the dynamic range.

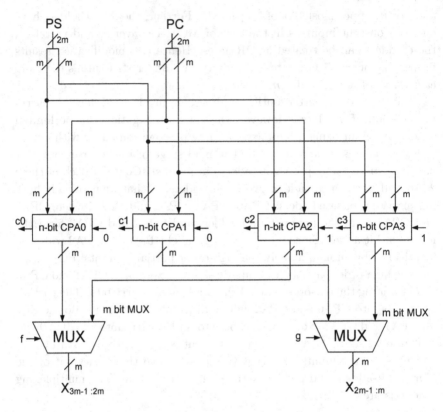

Figure 7.8: CPAs for $\{2^m - 1, 2^m, 2^m + 1\}$ reverse converter

7.5 High-radix conversion

If we wish to directly speed-up an implementation of the architecture of Figure 7.1, then it is evident that the depth of the modular addition tree should be reduced. The obvious way to do this is to compute with higher radices; that is, all input-operands should be in a large radix. High-radix computation is especially advantageous with architectures that rely primarily on look-up tables (e.g. ROMs), as there is then a minimization of the number of table look-up operations as well as the number of modular additions. Using a large radix also balances the size of the look-up tables, which in turn results in a more regular structure that is suitable for VLSI realization of the converter. Nevertheless, the selection of a large radix must be such that it reduces the complexity of both the initial and final radix conversions.

High-radix computation is traditionally achieved by taking several bits of an operand at a time: one for radix-2, two, for radix-4, three for radix-8, and, in general, r for radix-2^r. We next show how this may be combined with base-extension to achieve high-radix reverse conversion. In the first approach, the basic idea is to start by viewing the desired output in terms of a large radix. The second approach more closely corresponds to the outline in the introduction above.

Suppose the number X is represented by $\langle x_1, x_2, \ldots, x_N \rangle$, relative to the moduli $m_1, m_2, \ldots, m_N$, and that the representation of X requires m bits. Then X may be expressed as

$$X = \sum_{i=0}^{m-1} b_i 2^i$$

So, in order to determine X from its residues, we need to determine the bits $b_0, b_1, b_2, \ldots, b_{m-1}$. If the chosen radix is 2^r—i.e. operand-bits are taken r at a time—then, by grouping the b_is into k blocks, $\mathbf{b}_i$, of n bits each, X may be expressed as

$$X = \mathbf{b}_k 2^{(k-1)r} + \mathbf{b}_{k-1} 2^{(k-2)r} + \cdots + \mathbf{b}_2 2^r + \mathbf{b}_1$$

where the coefficients $\mathbf{b}_i$ are given by

$$\mathbf{b}_i = \sum_{i=0}^{n-1} 2^p b_{r(i-1)+p} \qquad 1 = 1, 2, \ldots k$$

Suppose r is chosen so that 2^r is larger than each of the other moduli, m_i, and base-extension is then carried out with the extra modulus $m_E = 2^r$. Then

$$\begin{aligned} x_E &\triangleq |X|_{m_E} \\ &= \left| \mathbf{b}_k 2^{(k-1)r} + \mathbf{b}_{k-1} 2^{(k-2)r} + \cdots + \mathbf{b}_2 2^r + \mathbf{b}_1 \right|_{2^r} \\ &= \left| \left| \mathbf{b}_k 2^{(k-1)r} \right|_{2^r} + \left| \mathbf{b}_{k-1} 2^{(k-2)r} \right|_{2^r} + \cdots + \left| \mathbf{b}_2 2^r \right|_{2^r} + \left| \mathbf{b}_1 \right|_{2^r} \right|_{2^r} \\ &= \mathbf{b}_1 \end{aligned}$$

Let us now define the integer Y_1 as

$$Y_1 = \frac{X - x_E}{2^r}$$

Then
$$2^r Y_1 = X - x_E$$
$$= \sum_{i=2}^{k} \mathbf{b}_i 2^{(i-1)r}$$

The number $2^r Y_1$ may then be expressed in terms of its residues as
$$2^r Y_1 = \langle |X - x_E|_{m_1}, |X - x_E|_{m_2}, \ldots |X - x_E|_{m_N} \rangle$$

The range of Y_1 is $[0, \lfloor (M-1)/2^r \rfloor)$. So it possible to express Y_1 uniquely, by multiplying each of the residues in the last equation by the multiplicative inverse of m_E:

$$Y_1 \cong \left\langle \left| \left| m_E^{-1} \right|_{m_1} (x_1 - x_E) \right|_{m_1}, \left| \left| m_E^{-1} \right|_{m_2} (x_2 - x_E) \right|_{m_2}, \ldots, \right.$$
$$\left. \left| \left| m_E^{-1} \right|_{m_N} (x_N - x_E) \right|_{m_N} \right\rangle$$

Since we have chosen m_E to be greater than each of the moduli m_j, the residues here may be dropped in the representation of Y.

If we now define Y_2 by
$$Y_2 = \frac{Y_1 - x_E}{2^r}$$

and a similar process of base-extension and scaling yields $\mathbf{b}_2$, $2^r Y_2 = \sum_{i=3}^{k} \mathbf{b}_i 2^{(i-1)r}$, and Y_2 is residue form. The scaling and base-extension processes are then repeated until all the $\mathbf{b}_i$s will have been determined. X is then obtained by adding up the $\mathbf{b}_i$s in a (shallow) tree of carry-save adders and a final carry-propagate adder. The values, $|X - x_E|_{m_1}, |X - x_E|_{m_2}, \ldots |X - x_E|_{m_N}$ are stored in a lookup-table and accessed as required.

In the preceding approach, the starting point was to imagine a high-radix representation of the desired output, X. A different approach would be to start "at the top" (of Figure 7.1). That is, instead of having ROMs that are each accessed with a single residue, each ROM will now be accessed with multiple residues [3]. Evidently, the trade-offs depend on the number, sizes, and speeds of the ROMs used in the realization. The basic idea here is as follows.

Let $S \triangleq \{m_1, m_2, \ldots m_N\}$ be a set of moduli and $P \triangleq \{s_1, s_2, \ldots s_l\}$ be a partition over S such that $s_i \subseteq S$, $s_i \cap s_j = \phi$ for $i \neq j$ and $S = s_i \cup s_j$. All residues corresponding to the subset P are looked-up once to produce a partial result $\widetilde{X}_p$ (Equation 7.2):

$$\widetilde{X}_p = \left| \sum_{m_i \in S} |x_i X_i|_{m_i} \right|_M$$

The sought result X is then obtained as

$$X = \left| \sum_{i=i1}^{l} \widetilde{X_p} \right|_M$$

Given that there are now fewer operands into the multi-operand modular adder (Figure 7.1), the tree will be shallower and faster. The only difficult part in applying this algorithm is how to determine an optimal partition; this requires an extensive search. Even the best search may result in imbalances in the residue channels. Fortunately, since combining does not have to be applied to all or entire residues, the combination may be such as to yield equal-sized blocks of bits of one or more residues, or it may be done on some other basis.

7.6 Summary

The main methods for reverse conversion are the Mixed-Radix Conversion (MRC) and application of the Chinese Remainder Theorem (CRT) . All other methods may be viewed as variants of these. For example, the Core Function, which at first glance appears to be a different approach, is in fact a variant of the CRT approach. The main drawback of the CRT is the requirement for modular reductions relative to large moduli. The MRC does not have this drawback, but it is inherently a sequential approach, unlike the CRT, for which non-trivial parallelism can be exploited. The Core Function does not have either of these drawbacks, but in practice it appears to be difficult to implement it with less hardware than in the CRT or MRC or to realize faster implementations. (This is an area that requires further research.) For the special moduli-sets and their extensions, reverse converters of good performance and cost are relatively easy to design, which explains the popularity of these moduli-sets.

References

(1) S. J. Piestrak. 1995. A high speed realization of a residue to binary number system converter. *IEEE Transactions on Circuits and Systems II*, 42(10):661–663.
(2) A. P. Shenoy and R. Kumaresan. 1989. Fast base extension using a redundant modulus in RNS. *IEEE Transactions on Computers*,

38(2):292–296.

(3) B. Parhami and C. Y. Hung. 1997. "Optimal Table Lookup Schemes for VLSI Implementation of Input/Output Conversions and other Residue Number Operations". In: *VLSI Signal Processing VII*, IEEE Press, New York.

(4) N. Burgess. 1999. "Efficient RNS to binary conversion using high-radix SRT division". In: *Proceedings, 14th International Symposium on Computer Arithmetic*.

(5) K. M. Elleithy and M. A. Bayoumi. 1992. Fast and flexible architecture for RNS arithmetic decoding. *IEEE Transactions on Circuits and Systems II*, 39(4):226–235.

(6) N. Burgess. 1997. "Scaled and unscaled residue number system to binary number conversion techniques using the core function". In: *Proceedings, 13th International Symposium on Computer Arithmetic*, pp 250257.

(7) A. B. Premkumar. 1992. An RNS binary to residue converter in the $2n-1, 2n, 2n+1$ moduli set. *IEEE Transactions on Circuits and Systems–II*, 39(7):480–482.

(8) M. A. Soderstrand, W. K. Jenkins, G. A. Jullien and F. J. Taylor. 1986. *Residue Number System Arithmetic: Modern Applications in Digital Signal Processing*, IEEE Press, New York.

(9) G. Bi and E. V. Jones. 1988. Fast conversion between binary and residue numbers. *IEE Letters*, 24(19):1195–1196.

(10) S. Andraos and H. Ahmad. 1988. A new efficient memoryless residue to binary converter. *IEEE Transactions on Circuits and Systems*, 35(11):1441–1444.

(11) S. Bandyopadhay, G. A. Jullien, and A. Sengupta. 1994. A fast systolic array for large modulus residue addition. *Journal of VLSI Signal Processing*, 8:305–318.

(12) S. J. Piestrak. 1994. Design of residue generators and multi-operand modular adders using carry-save adders. *IEEE Transactions on Computers*, (1):68–77.

(13) R. Conway. 1999. Fast converter for 3 moduli RNS using new property of CRT. *IEEE Transactions on Computers*, 48(8):852–860.

(14) M. Bhardwaj, A.B. Premkumar, and T. Srikanthan. 1998. Breaking the $2n$-bit carry propagation barrier in reverse conversion for the $\{2^n - 1, 2^n, 2^n + 1\}$ moduli-set. *IEEE Transactions on Circuits and Systems, I*, 45:998–1002.

(15) W. Lynch et al. 1998. "Low load latency through sum-addressed

memory". In: *Proceedings, 25th International Symposium on Computer Architecture*, pp 369–379.

Chapter 8

Applications

This chapter is a brief introduction to some applications of residue number systems. The two most significant properties of residue number systems are the absence of carry-propagation in addition and multiplication, and, consequently, the ability to isolate individual digits, which may be erroneous. Therefore, the most fruitful uses of these systems are likely to be in fault-tolerance and in applications in which addition and multiplication are the predominant arithmetic operations. A leading candidate for the latter is digital signal processing, which is covered in the first section of the chapter. Additionally, the carry-free nature of arithmetic operations facilitates the realization of low-power arithmetic, which is critical in many current systems and especially in embedded processors. Fault-tolerance, which is covered in the second section, is an area that was the subject of much research in the early days of computing, but such work subsequently declined when computer technology became more reliable. Now, with the advent of extremely dense computer chips that cannot be fully tested, fault tolerance and the general area of computational integrity have again become more important [20]. Another area of applications, that of communications, is also briefly considered, in the third section of the chapter.

In what follows, we shall briefly review the essentials of each of the three areas mentioned above; but an exhaustive exposition would be outside the scope of this book, and we shall therefore assume some relevant prior knowledge on the reader's part. As far as the use of residue number systems goes, limitations of space mean that this chapter should be regarded as no more than a very brief introduction.

8.1 Digital signal processing

Most "real-world signals are analog, and processing such signals to extract useful information is rather tedious. Moreover, this is complicated by the near-total absence of analog computers. Consequently, the digital processing of signals is the primary vehicle for the extraction of information. In digital signal processing (DSP), an input signal is converted into a sequence of numbers that are the result of sampling and quantizing analog signals. In the process of extracting useful information, the signals are usually transformed from one type of representation into another type in which certain characteristics of the signal become obvious. The main advantage that DSP offers is in the flexibility associated with the generation and modification of the algorithms used for these transformations. In addition, the digital processing of signals offers other advantages, such as easy programmability, stability, repeatability, and easy implementation of systems. By systems, we mean circuits that implement application algorithms in engineering fields, such as communication, image processing, encryption, data compression for transmission, and so forth. Digital processing also makes it easy to correct errors that occur during signal transmission or during data storage.

Signal analysis and application-algorithms in DSP are computationally demanding. The demand arises from the fact that the most frequently used arithmetic operation in almost all DSP algorithms is the sum-of-products, which consists of sequences of multiplication and addition. Multiplication can be an expensive operation, in terms of speed and chip-area, and can therefore constitute a significant bottleneck in the implementation of DSP algorithms. The multiply-accumulate operation should be very fast, and this is especially so in applications that require real-time processing, in which a particular concern is the amount of processing that can be done before new data arrives. Since multiplication is more time consuming than addition, if multiplication can be speeded up, then overall processing times can be brought down. The multiplication of large numbers is inevitably time-consuming, or costly in hardware, or both; so transforming data into the residue domain can effectively reduce multiplication times and costs: with a proper choice of moduli-sets, the multiplier and multiplicand can be represented in the residue domain [1], which in turn results in smaller and faster multipliers. In this chapter, we will discuss the implementation of DSP algorithms in this domain and consider the advantages and disadvantages of using residue numbers.

8.1.1 *Digital filters*

Digital filters are especially important in DSP because they can be used for a wide variety of applications: noise-reduction, band splitting, band limiting, interpolation, decimation, pulse forming, echo suppression, equalization, and so forth. An advantage of digital filters, in relation to their analog counterparts, is that of repeatability. This is because analog filters use components, such as resistors and capacitors, that have some associated tolerance. On the other hand, digital filters rely primarily on the sum-of-products operation, and, for the same set of operands, this yields the same result every time it is carried out. Also, a DSP system can, by a simple change of system parameters, be easily adapted to changes in the environments; this is not so with analog systems. Furthermore, DSP systems are reprogrammable, and so, for example, their characteristics from a low-pass filter to a high-pass or a band-pass filter can be changed without any physical changes.

Two basic types of filters are commonly implemented in DSP: Finite Impulse Response (FIR) filters and Infinite Impulse Response (IIR) filters. Before we proceed to discuss the implementation of these filters in the residue domain, we shall briefly introduce the underlying basic concepts.

8.1.1.1 *Finite Impulse Response filters*

The input to a digital filter is usually a sequence of numbers obtained from sampling an analog signal. The FIR filter may be represented by a block diagram of the type shown in the example of Figure 8.1. The z^{-1} block represents a *unit delay*. If the input is $x(n)$, then the output of the delay element is $x(n-1)$, which is the value of $x(n)$ one time-period before now, or, put simply, the previous input. Similarly, $x(n-2)$ simply means the value of the input two sampling periods before now. The filter shown in the Figure 8.1 may be represented by a difference equation:

$$y(n) = a_0 x(n) + a_1 x(n-1) + a_2 x(n-2)$$

The number of branches in the filter is known as the *taps* of the filter. The filter can be described in terms of its impulse response, which is a series of weighted impulses. With the impulse response, it is easy to compute the output of the filter simply by multiplying the impulse response and the input train of sampled pulses present at the desired time. We may use the idea of delay elements to rewrite the difference equation and so obtain the transfer function. Since z^{-1} stands for a delay element we have

$$y(n) = (a_0 + a_1 z^{-1} + a_2 z^{-2}) \times x(n)$$

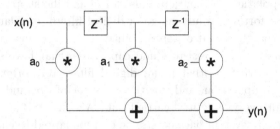

Figure 8.1: Three-tap FIR filter

The impulse response of the filter is

$$h(n) = \frac{y(n)}{x(n)}$$
$$= a_0 + a_1 z^{-1} + a_2 z^{-2}$$

In the case of the FIR filter, the impulse response has only a numerator polynomial, the roots of which are called *zeros* of the filter. The FIR filter is inherently stable due to the absence of denominator polynomial in its impulse response.

To see how easily an FIR filter can be implemented, consider an m-th order filter:

$$y(n) = \sum_{r=0}^{m} a_r x(n-r)$$

To understand the implementation better, we look at the second-order filter that we considered earlier. Here, $m = 2$, and the output at any time instant is given by the equation

$$y(n) = a_0 x(n) + a_1 x(n-1) + a_2 x(n-2)$$

For this three-tap filter, we require three multiplications and two additions. As indicated above, multiplication can be an expensive operation in terms of time, power, and area when realized on a VLSI chip. Nevertheless, if we can find a way to decompose large numbers into smaller numbers, perform multiplication on these smaller numbers in parallel, and some how combine the partial result to produce the same result as that of multiplying the two original numbers, then we can achieve significant saving in all of the above requirements. Such a scheme is relatively easy to implement in the residue domain: The operands are represented by their residues and multiplied together in parallel channels to obtain the residues of the product. A reverse conversion then gives the conventional equivalent of the product. In using RNS for FIR-filter implementation, it is advantageous to choose a moduli-set that comprises of a large number of small numbers. Then the residues and, hence, the width of the multipliers can be kept small. We will now consider the RNS implementation of a general N-tap FIR.

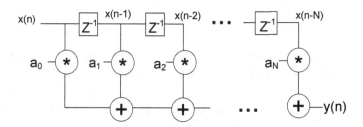

Figure 8.2: N-tap FIR filter

The output of an N-coefficient filter is given by the equation

$$y(n) = \sum_{i=0}^{N-1} a_i x_{n-i}, \quad n = 0, 1, \ldots, N-1$$

and the structure of the filter is shown in Figure 8.2

The general architecture for the RNS-based implementation is that shown in Figure 1.1. The moduli-set consists, of L small, relatively prime numbers, $m_1, m_2, \ldots, m_L$, whose product is sufficient to give the desired dynamic range. The FIR-filter implementation includes of a binary-to-residue converter at its input, to convert the input data into equivalent residues. The filtering is mainly performed in the central block. Since there are L residues in the residue set, L sub-filters are used to process corresponding residues from the input. The adders and multipliers may be any of the types described in Chapters 4 and 5, according to whatever trade-offs are sought. For better efficiency, the subfilters may be implemented as bit-serial units [2]; we discuss this next.

The input-output relationship with respect to a modulus m_j is given by the equation

$$y_{n,j} = \left| \sum_{i=0}^{N-1} a_{i,j} x_{n-i,j} \right|_{m_j}$$

where $a_{i,j} = |a_i|_{m_j}$, $x_{n-i,j} = |x(n-i)|_{m_j}$, $y_{n,j} = |y(n)|_{m_j}$, $n = 0, 1, \ldots,$ and $j = 1, 2, \ldots, L$. If we assume that m_j is represented in B bits, then

$$a_{i,j} = \sum_{b=0}^{B-1} a_{i,j}^b 2^b$$

$$x_{n-i,j} = \sum_{b=0}^{B-1} x_{n-i,j}^b 2^b$$

Here, $a_{i,j}^b$ and $x_{n-i,j}^b$ denote the b-th bits of the binary representation of $a_{i,j}$ and $x_{n-i,j}$ respectively. Now define, $S_{n,j}^b$ as

$$S_{n,j}^b = \left| \sum_{i=0}^{N-1} x_{n-i,j}^b a_{i,j} \right|_{m_j} \tag{8.1}$$

Then, the output, $y_{n,j}$ is given by the equation

$$y_{n,j} = \left| \sum_{b=0}^{B-1} S_{n,j}^b 2^b \right|_{m_j}$$

The output, $y_{n,j}$ is obtained by modulo-shifting and adding the elements of the sequence $S_{n,j}^b$. The FIR filter is realized once the outputs are determined. The efficiency in using RNS in FIR application depends on the computation of $S_{n,j}^b$. A simple algorithm to efficiently generate $S_{n,j}^b$ is given next.

The modular addition in the computation of $S_{n,j}^b$ can be performed quite simply as shown below. Let r_j denote the value $r_j = 2^B - m_j$, and let α_j and β_j be the residues of α and β respectively with respect to m_j, then the following theorem can be used in the computation of $S_{n,j}^b$.

THEOREM.

$$|\alpha_j + x \times \beta_j|_{m_j} = |(\alpha_j + x(\beta_j + r_j) + x \times c \times m_j)|_{2^B}$$

where c represents the complement of the carry bit generated during the operation $\alpha_j + x \times (\beta_j + r_j)$ and x is the input data to the filter. The proof is as follows [2].

Applications

The theorem is evidently true when x is 0. When x is 1, we need to consider two cases.

Case 1: $\alpha_j + \beta_j < m_j$. Then
$$\alpha_j + \beta_j + r_j < 2^B \quad \text{(implying } c = 1\text{)}$$
From the above equation, the carry generated from $\alpha_j + x \times (\beta_j + r_j)$ is 0. Since we have defined c as the complement of the carry, $c = 1$. So
$$|\alpha_j + x(\beta_j + r_j) + x \times c \times m_j|_{2^B} = |\alpha_j + \beta_j + r_j + m_j|_{2^B}$$
$$= |\alpha_j + \beta_j + 2^B|_{2^B}$$
$$= \alpha_j + \beta_j$$
$$= |\alpha_j + \beta_j|_{m_j}$$
$$= |\alpha_j + x \times \beta_j|_{m_j}$$

Case 2: $\alpha_j + \beta_j \geq m_j$. We have
$$2^B \leq \alpha_j + \beta_j + r_j < 2^{B+1} \quad \text{(implying } c = 0\text{)}$$
The carry generated from $\alpha_j + x \times (\beta_j + r_j)$ is 1; so $c = 0$, and
$$|\alpha_j + x(\beta_j + r_j) + x \times c \times m_j|_{2^B} = |\alpha_j + \beta_j + r_j|_{2^B}$$
$$= |\alpha_j + \beta_j + 2^B - m_j|_{2^B}$$
$$= \alpha_j + \beta_j - m_j$$
$$= |\alpha_j + \beta_j|_{m_j}$$
$$= |\alpha_j + x \times \beta_j|_{m_j}$$

END THEOREM

The result
$$|\alpha_j + x \times \beta_j|_{m_j} = |\alpha_j + x(\beta_j + r_j) + x \times c \times m_j|_{2^B}$$
may be used in the computation of $S^b_{n,j}$ in the following manner. Define
$$P^b_{n,j}(k) = \left| \sum_{i=0}^{k} x^b_{n-1,j} a_{i,j} \right|_{m_j}$$
This value can be computed iteratively, as follows.
$$P^b_{n,j}(-1) = 0$$
$$a'_{i,j} = a_{i,j} + r_j$$
$$P^b_{n,j}(0) = \left| P^b_{n,j}(-1) + x^b_{n,j} a_{0,j} \right|_{m_j}$$
$$= \left| P^b_{n,j}(-1) + x^b_{n,j} a'_{0,j} + x^b_{n,j} c^b_{n,j}(0) m_j \right|_{2^B}$$

where $c_{n,j}^b(0)$ is the complement of the carry bit generated in the addition of

$$P_{n,j}^b(-1) + x_{n,j}^b a'_{0,j}$$

Continuing the iterative process,

$$P_{n,j}^b(1) = \left| P_{n,j}^b(0) + x_{n-1,j}^b a_{1,j} \right|_{m_j}$$
$$= \left| P_{n,j}^b(0) + x_{n-1,j}^b a'_{1,j} + x_{n-1,j}^b c_{n,j}^b(1) m_j \right|_{2^B}$$

where $c_{n,j}^b(1)$ is the complement of the carry bit generated in the addition

$$P_{n,j}^b(0) + x_{n,j}^b a'_{0,j}$$

At the very last iteration, we obtain

$$P_{n,j}^b(N-1) = \left| P_{n,j}^b(N-2) + x_{n-N+1,j}^b a_{N-1,j} \right|_{m_j}$$
$$= \left| P_{n,j}^b(N-2) + x_{n-N+1,j}^b a'_{N-1,j} + x_{n-N+1,j}^b c_{n,j}^b(N-1) m_j \right|_{2^B}$$
$$= S_{n,j}^b$$

which is the desired output of the RNS FIR filter. subfilter.

Studies comparing RNS filters with conventional filters have shown that even with a fairly direct implementation, for the same performance, the RNS implementations have better area and power dissipation [26, 27].

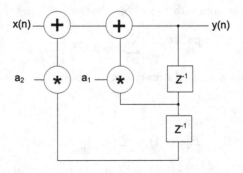

Figure 8.3: Structure of an IIR filter

8.1.1.2 Infinite Impulse Response Filters

A simple IIR filter has a structure of the type shown in the Figure 8.3. From this figure, we see that the output, $y(n)$ is given by the equation

$$y(n) = x(n) + a_1 y(n-1) + a_2 y(n-2)$$
$$= x(n) + \left[a_1 z^{-1} + a_2 z^{-2}\right] y(n)$$
$$= x(n) \frac{1}{1 - a_1 z^{-1} - a_2 z^{-2}}$$

For this filter, we can determine the transfer function:

$$H(z) = \frac{1}{1 - a_1 z^{-1} - a_2 z^{-2}} \qquad (8.2)$$

Observe that the output is dependent on all past inputs and keeps growing in length; that is, it is dependent on an infinite number of inputs, which is why the filter is called infinite impulse response filter. The structure filter has feedback loops that tend make it unstable. IIR filters have poor phase-response that are non-linear at the edge of the bands. Nevertheless, a merit of the IIR filter is that for the same stop-band attenuation, the number of taps required is smaller than for an FIR filter. There are several different forms for IIR filters, of which one of the common is the lattice structure.

In the implementation of digital filters, pipelining is often used to maximize throughput of the filter. Pipelining consists in placing delay units after each processing unit so that data can be processed in parallel for successive samples. This introduces a net delay, but that delay does not pose a problem in the case of FIR. In the case of IIR, however, this delay can cause problems, since the feedback units have to wait until the samples have been processed by the entire filter. Normally, a simple equation of the form shown in Equation 8.2 requires more than one delay. Using the Z transform, Equation 8.2 can be written as:

$$Y(z) = X(z) + a_1 z^{-1} Y(z) + a_2 z^{-2} Y(z) \qquad (8.3)$$

where $Y(z)$ is the Z transform of $y(n)$ and $X(z)$ is the Z transform of $x(n)$. So the filter requires that the hardware be able to add to the input, $x(n)$, the value of the output, $y(n)$, delayed by two periods. If, however, the pipelined architecture requires more than one delay, then Equation 8.3 cannot be realized. The equation can be modified to allow sufficient delay in the output without changing the resulting filter:

$$z^{-1} Y(z) = z^{-1} X(z) + z^{-2}(a_1 + a_2 z^{-1}) Y(z) \qquad (8.4)$$

We simplify Equation 8.3 by using Equation 8.4:

$$Y(z) = X(z) + a_1[z^{-1}X(z) + z^{-2}(a_1 + a_2z^{-1})Y(z)] + a_2z^{-2}Y(z)$$
$$= [1 + a_1z^{-1}]X(z) + z^{-2}[(a_1^2 + a_2) + a_1a_2z^{-1}]Y(z)$$

With this version, the hardware is allowed to delay Y by two periods before adding it to X. Thus, by a simple substitution, the delay allowed in the pipelined architecture is doubled. The delay can be further increased by reapplying the simple substitution. Although the resulting hardware will be of similar complexity, the new filter requires additional zeros and this must be realized by the zero-producing section. By re-arrangement of terms, we determine the new transfer function as

$$H(z) = \frac{(z + a_1)}{(z^2 - a_1z - a_2)(z + a_1)}$$

Introduction of additional poles is canceled by the zeros, and the existing poles must still be within the unit circle to guarantee the stability of the filter.

An RNS-based implementation of an IIR filter is quite straightforward, along similar lines as for the FIR filter.

8.1.2 *Sum-of-products evaluation*

When the number of fixed coefficients is bounded, it is possible to devise an efficient table-lookup implementation of a filter. The general techniques used in that case are known as *distributed arithmetic* (DA) [28]. DA uses block multiplies in arithmetic operations such as multiplications. Since DSP extensively uses convolution sums (multiply-and-accumulate operations) in many algorithms, DA is ideally suited for computing convolution sums that can be implemented with a table look-up. DA, by nature, is computationally efficient, and the advantages are evident in hardware implementations. DA is essentially a bit-serial process for the computation of an inner product, which means that the evaluation of the convolution sum is likely to be slow. Nevertheless, the apparent slowness of DA is not a disadvantage if the number of elements in each vector is nearly the same as the number of bits in each element. Modifications, such as pairing and partitioning of the input data are possible, and these make DA evaluation of convolution sums faster.

It is possible to exploit the advantages of both RNS and DA in the implementation of DSP algorithms. In this section, we describe a method for the modular application of DA in the computation of inner products.

This method is ideally suited for generating convolution sums, without the use of extensive memory whenever the number of coefficients is large, as is the case in, for example, radar signal processing. We shall first briefly describe inner-product generation using bit-serial techniques and then develop a theory for implementing the computation architecturally in the residue domain.

The inner product is the dot-product between two column vectors. So it requires multiplying numbers in pairs and adding the results. This can be performed in a bit-serial manner using DA principles [3]:

$$Y = \sum_{i=0}^{N-1} C_i X_i$$

$$= \sum_{i=0}^{N-1} C_i \sum_{j=0}^{L-1} x_{i,j} 2^j$$

where, C_i and X_i are the fixed coefficients and the input respectively and $X = \sum_{j=0}^{L-1} x_j 2^j$. Interchanging the order of summation,

$$Y = \sum_{j=0}^{L-1} 2^j \sum_{i=0}^{N-1} C_i x_{i,j}$$

Since $x_{i,j} \in [0, 1]$, it is easier to precompute the values $\sum_{i=0}^{N-1} C_i x_{i,j}$ and store them rather than compute them as and when $x_{i,j}$ arrive. This results in a memory requirement of 2^N locations for storing $\sum_{i=0}^{N-1} C_i x_{i,j}$ for different combinations of $x_{i,j}$. We can reduce memory requirements by recasting the input data as offset binary code instead of straight binary code, but this requires shifting of data for alignment. This approach requires a moderate amount of memory. In the case of integer data, partitioning the input and applying it in a parallel form results in reduced memory. Still, architectures that rely on data-partitioning in the memory require many adders in order to achieve substantial reductions in memory requirements. It is possible to modify the application of DA principles for computing the inner products, in such a way that the end results are reduced memory and simpler hardware that uses fewer adders.

The direct application of DA principles for the computation of the inner products is memory-intensive. The increase in memory is an exponential function of the number of coefficients, and data widths have a direct impact on the size of the memory. Consider the computation of the inner product $\sum_{i=0}^{N-1} C_i X_i$, where C_i are the N coefficients and X_i are the input data each

L bits wide. If M is an odd modulus, then the inner product in modular arithmetic is given by

$$Y = \left| \sum_{i=0}^{N-1} C_i X_i \right|_M$$

Now define $\zeta(g)$ as

$$\zeta_i(g) = \begin{cases} 1 \text{ if } |C_i|_M = g, \ g = 0, 1, 2, \ldots M-1 \\ 0 \text{ otherwise} \end{cases}$$

Using this definition, we may rewrite Y to combine X_is with same residues. That is, $|C_i|_M$ as a sum of M or fewer terms:

$$Y = \left| \sum_{g=0}^{M-1} g \sum_{i=0}^{N-1} \zeta_i(g) X_i \right|_M$$

$$= \left| \sum_{g=1}^{M-1} g \sum_{i=0}^{N-1} \zeta_i(g) X_i \right|_M$$

Considering all $|C_i|_M$ to be unique, this results in a sum of at most $M-1$ terms. Since X_i is L bits wide, this equation can be expanded into

$$Y = \left| \sum_{g=1}^{M-1} g \sum_{i=0}^{N-1} \sum_{j=0}^{L-1} 2^j \zeta_i(g) x_{i,j} \right|_M$$

where $x_{i,j}$ denotes the j-th bit of X_i. Interchanging the order of summation, as is done in DA, we have

$$Y = \left| \sum_{j=0}^{L-1} 2^j \sum_{g=1}^{M-1} \sum_{i=0}^{N-1} g \zeta_i(g) x_{i,j} \right|_M$$

$$= \left| \sum_{j=0}^{L-1} 2^j \sum_{i=0}^{N-1} \left[\sum_{g=1}^{(M-1)/2} g \zeta_i(g) x_{i,j} + \sum_{g=(M+1)/2}^{M-1} g \zeta_i(g) x_{i,j} \right] \right|_M \quad (8.5)$$

To proceed further, we will make use of the following result.

THEOREM.

$$|\beta X|_M = |\beta + (M - \beta)\overline{X}|_M, \quad X \epsilon(0,1)$$

Applying this theorem to the second term in the last equation and rewriting,

Applications

we have

$$\left| \sum_{g=(M+1)/2}^{M-1} g\zeta_i(g)x_{i,j} \right|_M = \left| \sum_{g=(M+1)/2}^{M-1} [g + (M-g)\bar{x}_{i,j}]\zeta_i(g) \right|_M$$

$$= \left| \sum_{g=(M+1)/2}^{M-1} g\zeta_i(g) + \sum_{g=(M+1)/2}^{M-1} (M-g)\bar{x}_{i,j}\zeta_i(g) \right|_M \quad (8.6)$$

Now define α_i as

$$\alpha_i = \sum_{g=(M+1)/2}^{M-1} g\zeta_i(g) \quad (8.7)$$

In order to simplify the architecture that is used to evaluate the sum, the second sum term in Equation 8.6 is rewritten by substituting $g = M - u$, where u is a dummy variable:

$$\left| \sum_{g=(M+1)/2}^{M-1} (M-g)\bar{x}_{i,j}\zeta_i(g) \right|_M = \left| \sum_{u=M-(M+1)/2}^{M-(M-1)} [M - (M-u)]\bar{x}_{i,j}\zeta_i(M-u) \right|_M$$

$$= \left| \sum_{u=(M-1)/2}^{1} u\bar{x}_{i,j}\zeta_i(M-u) \right|_M$$

$$= \left| \sum_{g=1}^{(M-1)/2} g\bar{x}_{i,j}\zeta_i(M-g) \right|_M \quad (8.8)$$

Using Equations 8.6, 8.7 and 8.8 in Equation 8.5, we have

$$Y = \left| \sum_{j=0}^{L-1} 2^j \sum_{i=0}^{N-1} \left[\alpha_i + \sum_{g=1}^{(M-1)/2} g\zeta_i(g)x_{i,j} + \sum_{g=1}^{(M-1)/2} g\bar{x}_{i,j}\zeta_i(M-g) \right] \right|_M$$

$$= \left| \sum_{j=0}^{L-1} 2^j \sum_{i=0}^{N-1} \left[\alpha_i + \sum_{g=1}^{(M-1)/2} g(\zeta_i(g)x_{i,j} + \bar{x}_{i,j}\zeta_i(M-g)) \right] \right|_M \quad (8.9)$$

Since modular multiplication principles are used in evaluating the sum of products, we write Equation 8.9 in a form that is appropriate for this operation:

$$Y = \left| \sum_{j=0}^{L-1} 2^j S_j \right|_M \quad (8.10)$$

where

$$S_j = \left| \sum_{i=0}^{N-1} \left[\alpha_i + \sum_{g=1}^{(M-1)/2} g(\zeta_i(g)x_{i,j} + \overline{x}_{i,j}\zeta_i(M-g)) \right] \right|_M \quad (8.11)$$

Equations 8.10 and 8.11 result in a low complexity and efficient hardware architecture. In arriving at an efficient architecture, we make use of the following theorem.

<u>THEOREM</u> Define $\overline{\gamma}$ to be the multiplicative inverse of γ with respect to M. Then $\overline{\gamma}$ exists if $\gcd(\gamma, M) = 1$, where $\gcd(\gamma, M)$ denotes the greatest common divisor of γ and M. Specifically, let $\gamma = 2$, then $\overline{2}$ exists, since M is odd. And

$$|\overline{2}S|_M = \begin{cases} \left|\frac{S+M}{2}\right|_M & \text{if } S \text{ is odd} \\ \left|\frac{S}{2}\right|_M & \text{otherwise} \end{cases}$$

This theorem is easily implemented in hardware using only two $(B+1)$-bit adders, where B is the number of bits used to represent M.

Expanding Equation 8.10 using the theorem gives

$$Y = \left|2^0 S_0 + 2^1 S_1 + 2^2 S_2 + \cdots + 2^{L-1} S_{L-1}\right|_M$$

$$= \left|\overline{2}^L 2^L S_0 + \overline{2}^{L-1} 2^L S_1 + \overline{2}^{L-2} 2^L S_2 + \cdots + \overline{2}^1 2^L S_{L-1}\right|_M$$

$$= \left|\left[\overline{2}^{L-1} 2^L S_0 + \overline{2}^{L-2} 2^L S_1\right]\overline{2} + \overline{2}^{L-2} 2^L S_2 + \cdots + \overline{2}^1 2^L S_{L-1}\right|_M$$

$$= \left|\left(\left[\overline{2}^{L-2} 2^L S_0 + \overline{2}^{L-3} 2^L S_1\right]\overline{2} + \overline{2}^{L-3} 2^L S_2\right)\overline{2} + \cdots + \overline{2}^1 2^L S_{L-1}\right|_M$$

$$\vdots$$

$$= \left|\left(\cdots\left(\left[\overline{2}2^L S_0 + 2^L S_1\right]\overline{2} + 2^L S_2\right)\overline{2} + \cdots + 2^L S_{L-1}\right)\overline{2}\right|_M \quad (8.12)$$

Observe from this last equation that Y can be computed by applying the theorem. To start, an input of $|2^L S_0|_M$ is applied, and the output $|\overline{2}2^L S_0|_M$ is obtained. In the next cycle, this output is added to a new input $|2^L S_1|_M$ and $|\overline{2}(\overline{2}2^L S_0 + 2^L S_1)|_M$ is computed and so on. The equation can be realized with the simple circuit shown in Figure 8.4.

Given the input $|2^L S_j|_M$ at the $(j+1)$st clock cycle, the output $\left|\sum_{j=0}^{L-1} 2^j S_j\right|_M$, of the circuit is obtained after L clock cycles. This output also corresponds to $|\sum C_i X_i|_M$, which is the required inner product.

The output of the circuit is bounded by $2M - 1$ after L clock cycles; so it may be necessary to subtract M once, if the result is not within the required range. Since the output is bounded by $2M - 1$ and is always less than 2^{B+1}, we can use an OR gate to compute the most significant bit of the output.

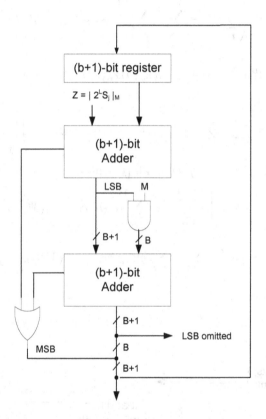

Figure 8.4: Distributed-arithmetic multiplier

In Equation 8.12, the input $|2^L S_j|_M$ is required for computing Y, where S_j is given by Equation 8.11 where S_j is given by Equation 8.11. The α term in S_j is added to the second term and this requires an extra adder. This addition and, therefore, the adder can be eliminated by simply absorbing the α_i term in the generation of $|2^L S_j|_M$, as follows.

$$|2^L S_j|_M = \left| 2^L \sum_{i=0}^{N-1} \left(\alpha_i + \sum_{g=1}^{(M-1)/2} g\left[\zeta_i(g)x_{i,j} + \zeta_i(M-g)\overline{x}_{i,j}\right] \right) \right|_M$$

$$= \left| 2^L \left(\sum_{i=0}^{N-1} \alpha_i + \sum_{i=0}^{N-1} \sum_{g=1}^{(M-1)/2} g[\zeta_i(g)x_{i,j} + \zeta_i(M-g)\overline{x}_{i,j}] \right) \right|_M$$

$$= \left| 2^L \left(\sum_{i=0}^{N-1} \alpha_i + \sum_{i=0}^{N-1} h\left[\zeta_i(h)x_{i,j} + \zeta_i(M-h)\overline{x}_{i,j}\right] \right. \right.$$

$$\left. \left. + \sum_{i=0}^{N-1} \sum_{g=1, g\neq h}^{(M-1)/2} g[\zeta_i(g)x_{i,j} + \zeta_i(M-g)\overline{x}_{i,j}] \right) \right|_M$$

where $h = 1, 2, \ldots, (M-2)/1)$ and $|\overline{h}|_M$ exists.

$$|2^L S_j|_M = \left| 2^L h \left[\overline{h} \sum_{i=0}^{N-1} \alpha_i + \sum_{i=0}^{N-1} (\zeta_i(h)x_{i,j} + \zeta_i(M-h)\overline{x}_{i,j}) \right] \right.$$

$$\left. + 2^L \sum_{g=1, g\neq h}^{(M-1)/2} \sum_{i=0}^{N-1} g[\zeta_i(g)x_{i,j} + \zeta_i(M-g)\overline{x}_{i,j}] \right|_M$$

(8.13)

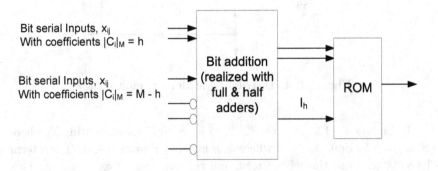

Figure 8.5: Circuit for Equation 8.14

Let $\delta = \overline{h}\sum_{i=0}^{N-1}\alpha_i$. Since the coefficients are fixed, δ can be precomputed, Equation 8.13 is rewritten using δ into

$$\left|2^L S_j\right|_M = \left|2^L h\left[\delta + \sum_{i=0}^{N-1}(\zeta_i(h)x_{i,j} + \zeta_i(M-h)\overline{x}_{i,j})\right]\right.$$
$$\left.+ 2^L \sum_{g=1,g\neq h}^{(M-1)/2}\sum_{i=0}^{N-1}g[\zeta_i(g)x_{i,j} + \zeta_i(M-g)\overline{x}_{i,j}]\right|_M \quad (8.14)$$

The circuit in Figure 8.5 implements first summation term, $|2^L h[\delta + \sum_{i=0}^{N-1}(\zeta_i(h)x_{i,j} + \zeta_i(M-h)\overline{x}_{i,j})]|_M$.

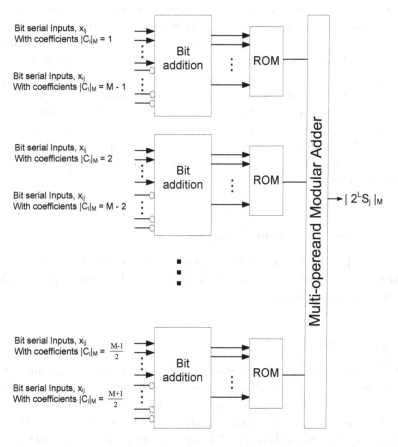

Figure 8.6: Circuit for evaluating $|2^L S_j|_M$

The contents of the ROM in the figure are the values $|2^L h(I_h + \delta)|_M$, I_h being $\sum_{i=0}^{N-1}(\zeta_i(h)x_{i,j} + \zeta_i(M-h)\overline{x}_{i,j})$. The second term of Equation 8.14 is implemented as a similar circuit whose ROM contents are given by $|2^L g(I_g)|_M$, I_g being $\sum_{i=0}^{N-1} g[\zeta_i(g)x_{i,j} + \zeta_i(M-g)\overline{x}_{i,j}]$. I_g and I_h are obtained by using either half or full adders.

Figure 8.6 shows how $|2^j S_j|_M$ is obtained from Equation 8.14, and this serves as the input to the modular multiplier shown in Figure 8.5. Observe that computation of $|2^L S_j|_M$, with the serial input $x_{i,j}$, consists of at most $(M-1)/2$ modulo adders, at most $2N$ memory locations, and a small number of 1-bit adders.

Extending RNS principles to the bit-serial mode of computation of inner products yields several advantages. First, no initial conversion of the input into residue domain is needed, although the computation is performed in the residue domain. This has been achieved by storing the products of the input bits and residues of the coefficients in memory. Second, the number of modulo adders is at most $(M-1)/2$, and this results in savings in area and power when the circuit is implemented in VLSI. Third, the proposed scheme is moduli-independent, since it requires only precomputing and storing data in ROM and additional adders. The memory required is a linear function of the number of coefficients, in contrast to the exponential function when direct DA principles are used. This results in reduced access times and lowers latency in the output. A final conversion is required to obtain the result in conventional form.

8.1.3 Discrete Fourier Transform

Before we proceed to the Discrete Fourier Transform (DFT), we will briefly introduce the Fourier Series which forms the basis for signal representation in the frequency domain.

8.1.3.1 Fourier Series

The basis functions that are used to represent any time-domain signal in the frequency domain are the trigonometric *sine* and *cosine* functions. Consider a rectangular pulse train in the time domain. This function consists of a large number of sine and cosines functions with varying amplitudes and frequencies. It is possible to describe any complex periodic signal as the sum of many sine and cosine functions. Fourier Series constitute one method that is commonly used to describe these periodic waveforms.

$$x(t) = \sum_{k=-N}^{N} C_k e^{j(k\omega_0 t)}$$

where C is a constant, $\omega_0 = 2\pi f_0$, and f_0 is the fundamental frequency. This equation can be used to represent any periodic waveform, provided there are a large number of terms in the summation. The multiple frequency terms appearing in the sum are called *harmonics*.

In discrete Fourier-Series representation, the continuous time equations are transformed into the discrete domain. This is done by replacing all functions involving the variable t by nT_s, where T_s is the sampling period:

$$x(n) = \sum_{-N}^{N} C_k e^{k\omega_0 nT_s}$$

In the above, we note that whenever

$$k\omega_0 T_s = 2\pi m$$

where m is an integer, the phase is indistinguishable. This means that the frequency response of a discrete signal is periodic, with a period $1/T_s$.

For non-periodic signals, we need to modify the Fourier Series. For a periodic signal, all frequencies are related. In the case of a non-periodic signal, we may say that $\omega_0 \to 0$. This implies that the number of basis functions tends towards infinity. In this case the summation tends to an integral:

$$x(t) = \frac{1}{2\pi} \int_{-\infty}^{+\infty} X(\omega) e^{j\omega t} d\omega$$

In this equation, we have assumed that the signal amplitude can be defined as function of frequency. The inverse Fourier Transform for the non-periodic signal then is

$$X(\omega) = \int_{-\infty}^{+\infty} x(t) e^{-j\omega t} dt$$

We now have a set of equations that is capable of representing the signal in both the frequency and time domains, and this set is called the Fourier-Transform pair. It is easy to see that the Fourier-Transform pairs are continuous and therefore cannot be implemented on any DSP chip or a digital computer. A discrete version of the transform is required if it is to be implemented on either a DSP chip or a digital computer. The discrete form can be obtained from the continuous version by simply replacing the continuous time variable t by the discrete variable nT_s. The discrete Fourier transform repeats with a period π/T_s. Changing the variable of integration, we get

$$x(n) = \frac{1}{2\pi} \int_{\pi}^{\pi} X(\omega) e^{j\omega n T_s} d(\omega T_s) \tag{8.15}$$

The reverse transform is then

$$X(\omega) = \sum_{n=-\infty}^{+\infty} x(n) e^{-j\omega n T_s} \tag{8.16}$$

The inverse transform uses summation instead of integration. Equations 8.15 and 8.16 allow us to transform discrete signals between time and frequency domains. We also observe that the frequency spectrum produced by the DFT is periodic, with a period ω_s. The DFT is important since it allows us to design IIR and FIR filters. These transforms can be used in many different ways, such as the computation of the frequency response of speech, image, and biomedical signals.

Applications involving periodic signals are rarely encountered. In most practical situations, we encounter aperiodic signals that are finite in length. The DFT of such a signal can be defined as

$$\begin{aligned} X(k) &= \sum_{n=0}^{N-1} x(n) e^{-j\frac{2\pi k n}{N}} \\ &= \sum_{n=0}^{N-1} x(n) W_N^{kn} \end{aligned} \tag{8.17}$$

where $W_N = e^{-j\frac{2\pi}{N}}$, and the spectral coefficients are evaluated for $0 \leq k \leq N-1$. The Inverse DFT (IDFT) that allows us to recover the signal is given by the equation

$$x(n) = \frac{1}{N} \sum_{k=0}^{N-1} x(n) W_N^{-kn} \tag{8.18}$$

In Equation 8.17, $x(n)$ is evaluated for $0 \leq n \leq (N-1)$. $X(k)$ is periodic beyond $N-1$, and so both the DFT and IDFT are finite-length sequences. Also, we note that the only difference between the DFT and IDFT is the scaling factor, N. So, if we have an algorithm for computing DFT, then we can use the same algorithm to compute the IDFT. This is because of the symmetry that exists between the time and frequency domains. The only difference between Equations 8.16 and 8.17 is that we have labeled the spectral coefficients as $X(k)$ instead of $X(\omega)$.

One of the major considerations in the computation of DFT is speed. Although, both the equations that are used for computing DFT and IDFT are straightforward, evaluating these equations takes time. Expressing the complex exponential using trigonometric functions, the DFT and IDFT equations become

$$X(k) = \sum_{n=1}^{N-1}\left[\cos(\frac{2\pi kn}{N}) - j\sin(\frac{2\pi kn}{N})\right]$$

and

$$X(n) = \frac{1}{N}\sum_{n=1}^{N-1}\left[\cos(\frac{2\pi kn}{N}) + j\sin(\frac{2\pi kn}{N})\right]$$

respectively. Direct implementation of these equations would be costly in terms of number of multiplications, since a total of $4N^2$ floating point multiplications is necessary. There are many efficient algorithms, called *FFT algorithms*, that use the symmetry of the transforms to reduce the complexity in the evaluation of the both DFT and IDFT.

8.1.4 RNS implementation of the DFT

The DFT is a versatile transform that is used in all most all engineering fields, for the analysis of complex waveforms, in which the DFT gives a clear indication of the different spectral components that make up the waveforms. In real-time application, requiring thousands of DFT transforms per second, a fixed-point solution must often be used in the design. Fixed-point arithmetic offers speed advantages, as long as the wordlength does not become overly large and high-speed multipliers are used. For near real time applications in signal analysis it is necessary to optimize the DFT to reduce both computation time and the hardware complexity of the DFT processor. As a solution, many implementations have turned to the RNS for two reasons: first, high precision can be achieved using several fast and simple processors; second, real-time programming and control tasks are simplified. The latter is an important strength of RNS. If arithmetic operations such additions and multiplications are implemented by lookup tables, then all these operations will be executed at the same speed. This eliminates the need for assuming worst-case delays when real-time code is prepared for executing DFT. The pipeline performance is then dependent on only the cycle access time. Further pipelining an RNS machine is straightforward because that data normally flows form one lookup table to its successor.

An optimized algorithm that implements the DFT is the FFT algorithm. The FFT algorithm exploits the symmetry that is present in the sample points in the original sequence when the number of samples is a power of two. Hardware realizations of the FFT range from the use of general purpose computers to specialized ROM-based implementations. For digital computers, in which finite register length is a constraint, the FFT cannot be implemented exactly. In the case of an integer implementation of FFT, there are two possible sources for errors: rounding or truncation of arithmetic results and the quantization errors introduced by the analog to digital converter. In applications where high precision has to be maintained, RNS is a viable alternative for small-wordlength architectures. To achieve the needed precision, one would only have to synthesize a large dynamic range by having a highly parallel system with datapaths of small wordlength. Nevertheless, RNS does present problems that are not normally found in weighted number systems; these include magnitude comparison and rounding.

The forward transform for a DFT is given by Equation 8.16, where N is a composite number. Therefore, the N-point DFT can be computed as a collection of smaller DFTs in conjunction with additional coefficients, sometimes called a *twiddle factors*. If the number of samples $N = r^m$, then the N is a composite number, with each factor equal to r. The algorithm used for computing the DFT in this case is referred to as an *radix-r* algorithm and may be implemented as a structure with m stages, where each stage has N/r basic r-point transforms. The basic form of the resultant r-point transform with a decimation-in-time (DIT) algorithm is given by the equation

$$X_{i+1}(k) = \sum_{n=0}^{r-1} x_i(n) (W_N)^n W_r^{nk}$$

where $x_i(n)$ denote the samples at the input of the i-th stage and $(W_N)^n$ are the corresponding twiddle factors.

In the RNS implementation of FFT, all non-integer coefficients, such as the twiddle factors, and all non-trivial intermediate results must be normalized to integers. All multiplication results are retained to full accuracy and the magnitudes of the numbers increase rapidly because of the cascaded multiplications. Since the range in RNS is the product of all moduli, the numbers must be scaled appropriately to be within the range of the RNS system to prevent overflows. Scaling operations are cumbersome in RNS; so it is necessary to maximize the ratio of binary operations to the numbers

Applications

of scaling operations. Since the magnitude of the numbers increase due to successive multiplications, it is advisable to choose a suitable radix that will keep the number of cascaded multiplications to a minimum. In using an FFT algorithm to compute the DFT, since the number of samples N is power of two, the FFT may be realized by using radices of 2^k, where k is a positive number. The number of cascaded multiplications for various r and m are shown in Table 8.1. This table has been generated on the assumption that there is only multiplication-level within the r-point DFT, for $r = 8$ and $r = 16$. For radices 2 and 4, the numbers shown in the table represent the cascaded twiddle factors only; for radices 8 and 16, the numbers represent the sum of cascaded twiddle factors and cascaded internal multiplications in the r-point DFTs.

Table 8.1: Cascaded multiplications for FFT

r	\multicolumn{6}{c}{N}
	64	128	256	512	1024	2048
2	4	5	6	7	8	9
4	2		3		4	
8	3			5		
16			3			

Table 8.2: Throughput vs. complexity of FFT implementation

Delay Δns	Radix-4 CRNS Butterfly			
	Maximum number of			
	Multipliers	Adders	Subtractors	Levels
1	12	11	11	4
2	6	6	6	4
4	3	3	3	4
8	2	2	2	4

Table 8.1 shows that radices 4 and 16 have fewer cascaded multiplications. This means that use of these radices will result in fewer scaling operations. It has been established that radix-4 FFT is an optimal realiza-

tion of DFT using decimation in time. Figure 8.7 shows the architecture for radix-4 complex implementation of the FFT algorithm. This consists of several complex-RNS units, adders, and subtractors. (Complex-RNS arithmetic is discussed in Chapter 2.)

It is possible to have a variety of throughput complexity tradeoffs in the above architecture. The throughput delay can be minimized by having concurrent hardware in the above architecture. If we consider Δ to be the worst case delay for an L moduli RNS design, the tradeoff between throughput and complexity can be summarized as shown in Table 8.2.

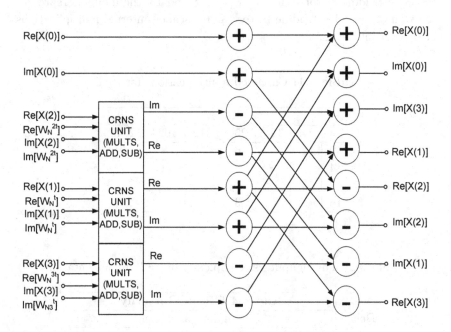

Figure 8.7: Architecture for radix-4 FFT

8.2 Fault-tolerance

As computer chips become increasingly dense, the probability that some part may fail also increases; however, the large number of components also means that exhaustive testing is not likely to be practical. Therefore, computational integrity has again become critical [20]. In this section we discuss issues relating to computer reliability and how to achieve fault-free functioning of computer systems [7,8,9]. Residue number systems can be useful in that regard.

There are two possible ways in which computers can be made more reliable. The first is to make sure that the computing system remains free of faults, which simply means that any potential fault that could arise during the operation of the computer is detected and eliminated before reliability of the system is affected. This also requires that no new faults develop; so the approach is usually referred to as *fault-prevention*. The second, and more practical, way is to accept no system is perfect and to include in the system measures for dealing with faults as they arise. This is called *fault-tolerance*. Fault-prevention requires that we be able to foresee all possible faults that may arise in the system—a formidable task, since it is difficult to predict the age of the physical components that are used in the design. We may, therefore, conclude that it is difficult to ensure high reliability through fault-prevention and that fault-tolerance is preferable.

The subject of fault-tolerance is vast, and, therefore, we shall restrict ourselves to one specific area in fault-tolerance: how prevent errors that occur in a system. Here, it should be noted that there is a difference between a fault and an error. Whenever there is a fault, the fault may cause an error in the system. This error can subsequently result in the failure of the system. Thus, the aim of the fault-tolerance is to prevent errors that may lead to system failure.

There are different phases associated with fault-tolerance: error-detection, error-recovery and fault-treatment. Error detection is necessary in order to tolerate faults. Although faults arising in a system cannot be detected directly, errors arising out of faults can be detected. The damage from error may lead to an erroneous system state. Error-recovery techniques transform an erroneous system state into a well-defined state that allows treatment to be effected in the subsequent state. Fault-treatment enables a system to continue providing service, by guaranteeing that the fault in question does recur immediately. It should be noted that the detection of an error is not indicative of the type of fault that has occurred. For example, an error in parity does not necessarily mean a transmission error. Therefore, the first step in fault-treatment is to identify the nature of a fault and then repair the fault or reconfigure the system so as to avoid the fault.

A fundamental concept in fault-tolerance is that of namely *redundancy* Redundancy implies replication, and fault-tolerance depends on the effective deployment and utilization of the redundant hardware. Redundancy may be implemented in hardware or in software. Hardware redundancy can be categorized as *static* or *dynamic*. The difference between the two types of redundancy lies in how redundant components are used. In static redundancy, the redundant components within the system function in such a way that effects from failure are not perceived by the system, whereas dynamic redundancy provides an error-detection capability within the system, and redundancy must be provided elsewhere to insure fault-tolerance.

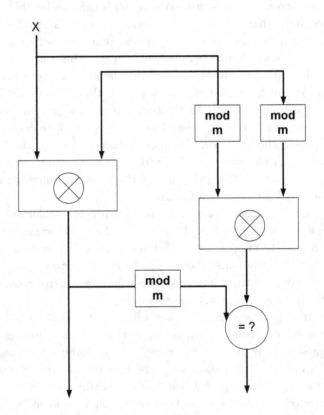

Figure 8.8: Error-checking using RNS codes

Although the primary motivation in the study of residue number systems has more been the implementation of fast arithmetic operations, there are some interesting properties associated with these systems that make them ideally suited for the detection and correction of errors. The relevant number systems here are the redundant ones that (RRNS) have been introduced in Chapter 2. Regardless of whether RRNS are used, RNS codes, because of their relatively small size, can facilitate the implementation inexpensive error-checking through redundancy. For example, a simple means to check the result of an arithmetic operation would be to duplicate the arithmetic unit, concurrently carry out the operation in two processors, and then compare the two results for equality. As an alternative, using RNS, we may proceed as follows. Instead of full replication of the arithmetic unit, an RNS unit is used, the advantage being that for a suitable modulus, the RNS processor will generally be smaller than the corresponding conventional unit. Then, for the arithmetic operation, $X \otimes Y$, we concurrently compute both $Z = X \otimes Y$ and $Z_{RNS} = |X_{RNS} \otimes Y_{RNS}|_m$, where m is some modulus, $X_{RNS} = |X|_m$, and $Y_{RNS} = |Y|_m$. And from Z, we compute $Z^*_{RNS} = |Z|_m$. Evidently, if no error has occurred, then we should have $Z^*_{RNS} = Z_{RNS}$. The corresponding hardware organization is therefore as shown in Figure 8.8. To ensure that the implemented hardware is efficient as possible, some care is needed in the choice of m; from preceding discussions, a modulus of the form $2^n - 1$ is a good choice.

The arrangement of Figure 8.8 has several obvious limitations. One is that, as the results of Chapter 2 show, it is most useful only for operations such as addition, subtraction, and multiplication. The other is that although we can determine whether or not an error has occurred, we have no means for locating the error or of correcting it, both of which can be accomplished if *all* operations are carried out in the RNS domain.

Residue number systems, through the use of redundant moduli allow the detection and correction of errors. Here, there are two essential aspects of RNS. The first is that because the residue representations carry no weight information, an error in any digit-position in a given representation does not affect any other digit-positions. The second is that there is no significant-ordering of digits in an RNS representation, which means that faulty digit-positions may be discarded with no effect other than the reduction of dynamic range. So, provided computed values can be constrained to the new dynamic range, obtained by omitting the faulty digit-position, computation can continue. These two properties therefore provide a basis for a sort of fail-soft capability. Furthermore, if enough redundant moduli are employed, then errors can be located precisely and corrections made. In what follows, we shall initially assume only single-digit errors. There basic results for such cases are from [7, 8, 14, 17]. Before proceeding, the reader may find it the discussion in Chapter 1 of RRNS.

An RRNS with a single redundant modulus is sufficient for the detection of single-digit errors. In such a system, it can be shown that if such an error has occurred, then the number represented by the erroneous residue representation will lie in the illegitimate range of the RRNS.

EXAMPLE. Take the RRNS with the non-redundant moduli 2,3, and 5 and the redundant modulus 7. The legitimate range is $[0, 30)$, and the total range is $[0, 210)$. In this system $23 \cong \langle 1, 2, 3, 2 \rangle$. Suppose a one-bit error occurs in the third residue, thus changing it from 3 to 2; that is, we have the erroneous value $\widetilde{X} \cong \langle 1, 2, 2, 2 \rangle$. The value of $\widetilde{X}$ is 127, which lies in the illegitimate range. END EXAMPLE

Therefore, the detection of single-digit errors can be accomplished by the application of the Chinese Remainder Theorem (CRT) or Mixed-Radix Conversion (MRC), to convert from residue representations, followed by a range-determining comparison. It should, however, be noted that here full mixed-radix conversion is not required: If no error has occurred, then, because the result will be in the legitimate range, the digits corresponding to the redundant moduli must all be 0s [21]. Therefore, an error is known to have occurred if at least one of those digits is non-zero. In other words, it is sufficient to determine only the mixed-radix digits; the actual value that they represent is not required.

In order to be able to both detect and correct single-digit errors, at least two redundant moduli are required. Evidently, the critical issues here are how to isolate a faulty digit and how to determine its correct value. There are two general ways in which these can be done. The first relies on base extension, via the CRT or MRC; the second is based on the concept of *projections* and is a variant of the use of the MRC [7].

Base extension may be used as follows [17]. During forward conversion, the values of the redundant residues are computed. An arithmetic operation is then carried out on the result-residues and used to obtain the redundant residues of the result. If the two sets of redundant residues do not match, then an error has occurred. The corrections can be simply implemented by using the differences between the two residue-sets to address a lookup table that yields the correct residues; this arrangement is shown in Figure 8.9. Alternatively the lookup tables may be replaced with additional computations. It should also be noted that if the moduli used are of the special types, then base extension can be realized easily.

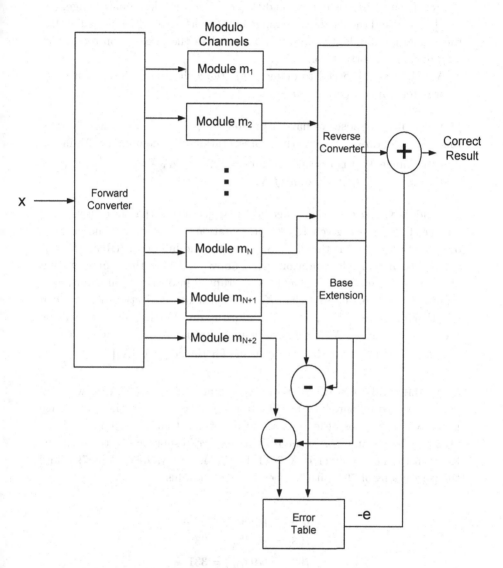

Figure 8.9: Error correction using through base extension

In what follows, we shall assume that we have a redundant residue number system (RRNS) in which $m_1, m_2, \ldots, m_N$ are the non-redundant moduli and $m_{N+1}, m_{N+2}, \ldots, m_{N+R}$ are the redundant moduli. That is, the total range, M_T, is determined by $\prod_{i=1}^{N+R} m_i$, the legitimate range, M, by $\prod_{i=1}^{N} m_i$, and the illegitimate range by $M_T - M$. We will also assume that the smallest redundant modulus is larger than the largest non-redundant modulus, an important condition.

We shall now proceed to define *projections* show how they are used in error detection and correction.

DEFINITION. Suppose we have a moduli-set $\{m_1, m_2, \ldots m_K\}$ and that in this system $\langle x_1, x_2, \cdots x_K \rangle$ is the representation of some number, X, in the dynamic range; that is $x_i = |X|_{m_i}$, for $i = 1, 2, \ldots, K$. Then, $X_i \stackrel{\triangle}{=} |X|_{M/m_i}$ is said to be the m_i-*projection of* X.

Evidently, the essential aspect of a projection is that it leaves out a residue digit from a given RNS-representation. What makes the concept of projections useful is this [21]: Suppose the moduli of an RRNS ($R = 2$) satisfy the magnitude-condition given above and that the representation of a legitimate number, X, has been corrupted into that of an illegitimate number, $\widetilde{X}$. Also, suppose that $\widetilde{X}_i$ is a legitimate projection of $\widetilde{X}$. Then all other projections, $\widetilde{X}_j$, $j \neq i$ are illegitimate. Furthermore, $\widetilde{X}_i$ will be the correct value for X. Therefore, it must be the case that the error is in the residue x_i and that the correct value for that digit is $\left|\widetilde{X}_i\right|_{m_i}$.

EXAMPLE. Take the RRNS with the moduli-set $\{2, 3, 5, 7, 11\}$, where 7 and 11 are redundant. The legitimate range is $[0, 30)$, and the total range is $[0, 2310)$. Suppose the representation of the legitimate number $23 \cong \langle 1, 2, 3, 2, 1 \rangle$ has an error in the residue-digit corresponding to the modulus 3. Then the new number is $\widetilde{X} \cong \langle 1, 1, 3, 2, 1 \rangle$. The value of $\widetilde{X}$ is 793, and the projections of 793 with respect to each modulus are

$$793_2 = |793|_{1155} = 793$$
$$793_3 = |793|_{770} = 23$$
$$793_5 = |793|_{462} = 331$$
$$793_7 = |793|_{330} = 133$$
$$793_{11} = |793|_{210} = 163$$

Observe that now there is only one legitimate projection—the one with respect to the modulus 3. We therefore conclude that the erroneous digit is in the second residue and that the correct value for that digit is $|23|_3$, which is 2. END EXAMPLE

In general, at least two redundant moduli are necessary for correction when exactly one residue is in error.

EXAMPLE. Consider the last example but without the redundant modulus 11—that is, the total range is now $[0, 210)$—and suppose that $\langle 1, 2, 3, 2\rangle$ (the representation of 23) has been corrupted into $\langle 1, 1, 3, 2\rangle$. The projections are now

$$793_2 = |793|_{105} = 58$$
$$793_3 = |793|_{70} = 23$$
$$793_5 = |793|_{42} = 37$$
$$793_7 = |793|_{30} = 13$$

We now have two legitimate projections but cannot tell which corresponds to the erroneous digit. END EXAMPLE

From the above, it would appear that in an RRNS with a total of $N + R$ moduli, $N + R - 1$ projections are required in order to isolate a single faulty digit. Nevertheless, the results above are readily extendible to the case where a projection is defined with respect to several moduli [15]. Thus the number of projections can be reduced by taking several moduli at a time.

We may now generalize our observation as follows. Suppose we have an RRNS with R redundant moduli, such that $R \geq 2$ and the moduli satisfy the magnitude-condition stated above. Then a single residue-error in the i-th position in the representation of X, a number in the legitimate range, can be unambiguously located. Furthermore, there will be only one m_i-projection of the resulting illegitimate number that lies in the legitimate range, and this number is the correct, legitimate value.

In general, the above observation can be extended, to this: if there are R redundant moduli, then it is possible to detect $R - 1$ errors and correct $R/2$ errors.

The methods above that rely on MRC are inherently slow, because the MRC is a sequential process. The CRT, which provides a faster (in theory) method for reverse conversion and base extension, may be used instead. For the detection of single-digit errors, it can be shown that approximate CRT decoding, rather than full decoding, is sufficient [11]; such decoding is based on the scaling of values by a power of two, as discussed in Chapters 2 and 6. Error-correction must still be carried out as above, e.g. via projections.

8.3 Communications

The multiple-access technique used in communication engineering is one way of efficiently allocating a rare communication resource, namely, the radio spectrum. This techniques becomes meaningful when a large number of users seek to communicate with each other simultaneously. This sharing of the frequency spectrum must be done in such a way that it does not negatively affect the quality of performance of the system. It is also often desirable to allow a user to send and receive information simultaneously. This can be achieved by duplexing in the time domain or in the frequency domain. Multiple-access techniques allow several users to communicate simultaneously while providing each user with duplexing capability.

There are three major access techniques that are used to share available bandwidth: *frequency-division multiple access* (FDMA), *time-division multiple access* (TDMA), and *code-division multiple access* (CDMA) These three techniques can be broadly categorized as *narrowband* and *wideband*.

In narrowband FDMA, the available radio spectrum is divided into a large number of narrowband channels, and each user is allocated a frequency band. Duplexing in this scheme is achieved by employing two frequencies within each band. In order to minimize interference between the transmitted and received messages for the same user, a large frequency-separation is imposed between the two. In narrowband TDMA systems, the same frequency is used for many users; but each user is allocated a unique time-slot in a cyclic manner, and within that time-slot, the user is allowed to transmit and receive information. Duplexing in TDMA is achieved by either frequency-division or time-division.

CDMA is a form of *spread-spectrum* multiple-access in which the transmission bandwidth is several orders of magnitude greater than the minimum required RF bandwidth. In this scheme, a narrowband signal is translated into a wideband noise-like signal before being transmitted. In general, spread spectrum techniques are robust against interference. The spread-spectrum technique is not spectrally efficient if the entire bandwidth is used for a single user. Nevertheless, by allowing several users to access the same channel and separating each from the other, spread-spectrum techniques can be made efficient. The most commonly used spread-spectrum techniques are *frequency hopped multiple access* (FHMA) and CDMA. In FHMA the carrier frequency of individual users is varied in a pseudo-random manner. Multiple access is achieved by allowing each user to dwell at a specific narrowband channel for a particular period of time. The duration and the narrowband channel are determined by a pseudo-random code allocated to that user. Information is divided into uniformly sized packets that are transmitted on different channels. The difference between FDMA and FHMA systems lies in the fact that in FHMA system, the frequency-hopped signal from the user changes channels at a rapid rate, whereas in the FDMA system the frequency allocated to a single user remains fixed during the entire usage period.

In CDMA systems, the message is *exclusively-ored* with a very large bandwidth signal called the *spreading signal*. The spreading signal is usually a pseudo-random code sequence that consists of a large number pulses called *chips*. In CDMA, several users access the same channel, and separation between users is accomplished by assigning each user a unique spreading code. In the receiver, the signals are time-correlated with the desired user code [13].

CDMA techniques differ from FDMA and TDMA in several aspects:

- In a duplex environment, the transmission techniques used in both directions are different, unlike in FDMA and TDMA;
- in CDMA, there are two stages of modulation;
- variable bit rate is used to achieve higher efficiency;
- and each radio channel is used only once whereas in FDMA and TDMA each user is allocated a channel.

In order to minimize interference, the pseudorandom codes that are assigned to different users have to be unique. Therefore, if there are a large number of users, the number of such sequences to be generated is also large. The use of RNS in generating the CDMA signals is explained next.

Consider a typical *M-ary* orthogonal CDMA system. In the physical layer of the reverse channel, the input data is first coded using convolution coding techniques. In order to maintain the required data rate, the coded bits are repeated and then interleaved to prevent narrowband interference from corrupting the data. The data bits are grouped into blocks of, say, 6 bits each. Each block of 6 input bits produces a sequence of 64 bits. These 64 bits correspond to a row in a 64×64 Walsh modulator. Each row in a Walsh matrix is orthogonal to every other 63 rows. By encoding 6-bit blocks of the Walsh matrix, the symbols are made orthogonal. The output from the Walsh encoder is the modulated radio-frequency carrier.

In order to see how RNS can benefit the CDMA system, in terms of the number of orthogonal sequences, consider a similar conventional system. Let $\{7, 11, 13, 15, 16\}$ be the moduli-set chosen for the RNS. The dynamic range is given by $\prod_{i=1}^{5} m_i - 1 = 240239$ The symbol length is then $\left\lfloor \log_2 \left(\prod_{i=1}^{5} m_i \right) \right\rfloor = 17$. So, any symbol of 17 bits can be chosen within the given dynamic range. For any number within the dynamic range, each of the moduli, $m_i, 1 \leq i \leq 15$, will at most produce m_i residues. Therefore, the total of all possible residues that a given number will produce is $\sum_{i=1}^{5} m_i = 62$. In this RNS-based CDMA system, only 62 orthogonal signals are required. So, to transmit a 17-bit symbol, only 5 residues are required, and five orthogonal waveforms corresponding to these five residues can be chosen from a set of 62 orthogonal signals. On the other hand, in the conventional M-ary orthogonal CDMA systems, the number orthogonal signals required will then be $2^{17} = 240240$, several orders of magnitude larger than that of the RNS-based CDMA system.

In summary, there are two other primary advantages to using RNS in the CDMA systems. First, with extra moduli, fault detection and correction features can readily be incorporated into the system. Second, since the number of bits per symbol is increased, the data rates can also be increased. A description and performance analysis of an RNS CDMA system will be found in [13].

8.4 Summary

Residue number systems are well suited for applications in which the predominant arithmetic operations are addition and multiplication. This includes many algorithms used in digital signal processing, and we have discussed the particular cases of digital filters and transforms. Residue number systems can also be usefully applied in the implementations of communication systems; a very brief example has been given.

The detection and correction of errors will become increasingly important as computer chips become larger and more dense. Residue number systems have properties—the absence of carries in arithmetic operations and the lack of significance-ordering among digits of a representation—that make them potentially very useful in this regard. We have given a brief introduction of such matters. In that introduction, we have implicitly assumed that we are dealing with fixed-point arithmetic. Nevertheless, residue number systems can also be useful for error detection and correction in floating-point arithmetic [16, 25]. Examples of other applications of RNS will be found in [29, 30, 31, 32].

References

(1) W. K. Jenkins. 1980. Complex residue number arithmetic for high speed signal processing. *Electronic Letters*, 16(17):282–283.

(2) C. L. Wang. 1994. New bit serial VLSI implementation of RNS FIR digital filters. *IEEE Transactions on Circuits and Systems–II*, 41(11): 768–772

(3) K. P. Lim and A. B. Premkumar. 1999. A modular approach to the computation of the convolution sum using distributed arithmetic principles. *IEEE Transactions on Circuits and Systems–II*, 46(1):92–96.

(4) F. J. Taylor. 1990. An RNS Discrete Fourier Transform implementation. *IEEE Transactions on Acoustics, Speech and Signal Processing*, 38(8):1386–1394.

(5) B. Tseng, G. A. Jullien and W. C. Miller. 1992. Implementation of FFT structures using the residue number systems. *IEEE Transactions on Computers*, 28(11):1453–1979.

(6) F. J. Taylor. 1987. A residue arithmetic implementation of FFT. *Journal of Parallel and Distributed Computing*, 4:191–208

(7) F. Barsi and P. Maestrini. 1973. Error correcting properties of redundant residue number systems. *IEEE Transactions on Computers*, 22(2):307–315.

(8) Barsi and P. Maestrini. 1974. Error detection and correction by product codes in residue number systems, *IEEE Transactions on Computers*, 23(9):915–923.

(9) V. Ramachandran. 1993. Single error correction in residue number system. *IEEE Transactions on Computers*, 32(3): 504–507.

(10) R. J. Cosentino. 1988. Fault-tolerance in systolic residue arithmetic processor array. *IEEE Transactions on Computers*, 37(7): 886–890.

(11) G. A. Orton, L. E. Peppard, and S. E. Tavares. 1992. New fault-

tolerant techniques for residue number systems. *IEEE Transactions on Computers*, 41(11): 1453–1464.
(12) E. D. Di Claudio, G. Orlandi and F. Piazza. 1993. A systolic redundant residue arithmetic error correction circuit. *IEEE Transactions on Computers*, 42(4): 427–432.
(13) L. L. Yang and L. Hanzo. 1999. Residue number system based multiple code DS-CDMA systems. In: *Proceedings, International Symposium on Circuits and Systems (ISCAS)*, pp: 1450–1453.
(14) D. Mandelbaum. 1972. Error correction in residue arithmetic. *IEEE Transactions on Computers*, C-21:538–545.
(15) S.-S. Wang and M.-Y. Shen. 1995. Single error residue correction based on k-term m_j projection. *IEEE Transactions on Computers*, 44(1):129–131.
(16) J.-C. Lo. 1994. Reliable floating-point arithmetic algorithms for error-coded operands. *IEEE Transactions on Computers*, 43(4):400–412.
(17) R. W. Watson and C. W. Hastings. 1966. Self-checked computation using residue arithmetic. *Proceedings of the IEEE*, 54:1920–1931.
(18) S. S. Yau and Y. Liu. 1973. Error correction in redundant residue number systems. *IEEE Transactions on Computers*, C-22:5–11.
(19) C.-C. Su and H.-Y. Lo. 1990. An algorithm for scaling and single residue error correction in residue number systems. *IEEE Transactions on Computers*, 39(8):1053–1063.
(20) M. J. Flynn and P. Huang. 2005. Microprocessor design: thoughts on the road ahead. *IEEE Micro*, 25(3):16–31.
(21) M. H. Etzel and W.K .Jenkins. 1980. Residue number systems for error detection and correction. *IEEE Transactions on Acoustics, Speech, and Signal Processing*, ASSP-28:538–544.
(22) F. Barsi and M. Maestrini. 1978. A class of multiple-error-correcting arithmetic residue codes. *Information and Control*, 36:28–41.
(23) W. K. Jenkins and E. J. Altman. 1988. Self-checking properties or residue number error checkers based on mixed radix conversion. *IEEE Transactions on Circuits and Systems*, 35:159–167.
(24) W. K. Jenkins. 1983. Design of error checkers for self-checking residue number arithmetic. *IEEE Transactions on Computers*, C-32:388–396.
(25) E. Kinoshita and K.-J. Lee. 1997. A residue unit for reliable scientific computation. *IEEE Transactions on Computers*, 46(2):129–138.
(26) GC. Cardarilli, A. Nannarelli and M. Re. 2000. Reducing Power Dissipation in FIR Filters using the Residue Number System, in: *Proceedings, 43rd IEEE Midwest Symposium on Circuits and Systems*.

(27) A. Nannarelli, M. Re and GC. Cardarilli. 2001. in: *Proceedings, International Symposium on Circuits and Systems (ISCAS)*, Vol. II, pp. 305–308
(28) C. S. Burrus. 1977. Digital filter structures described by distributed arithmetic. *IEEE Transactions on Circuits and Systems*,
(29) W. Wang, M. N. S. Swamy, and M. O. Ahmad. 2004. RNS applications for digital signal processing. In: *Proceedings, 4th International Workshop on System-on-Chip Real-Time Applications (IWSOC'04)*.
(30) S.-M. Yen, S. Kim, S. Lim, and S.-J. Moon. 2003. RSA speedup with Chinese Remainder Theorem immune against hardware fault cryptanalysis. *IEEE Transactions on Computers*, 52(4):461–472.
(31) Javier Ramrez 1, Antonio Garca 1, Pedro G. Fernndez 2 and Antonio Lloris. 2000. An efficient RNS architecture for the computation of dicrete wavelet transform on programmable devices. In: *Proccedings, X European Signal Processing Processing Conference*, pp. 255–258.
(32) A. Drolshagen, C. Chandra Sekhar, and W. Anheier. 1997. A residue number arithmetic based circuit for pipelined computation of autocorrelation coefficients of speech signal. In: *Proceedings, 11th Conference on VLSI Design: VLSI for Signal Processing*.

Index

adder
  carry-lookahead, 91
  carry-ripple, 85
  carry-save, 141
  carry-select, 108
  carry-skip, 88
  conditional-sum, 97
  multi-operand, 63
  multi-operand modular, 63
  parallel-prefix, 127
  ripple, 85
addition
  residues, 29
additive inverse, 27
arbitrary moduli-sets
  conversion, 68
  forward conversion, 58
assimilation, 141
base, 6
base extension, 14, 223
basic period, 66
Booth recoding, 144
carry
  generation, 91
  partial, 141
  propagate, 88
  propagation, 91
  transfer, 91
carry-lookahead adder, 91
carry-ripple adder, 85
carry-save adder, 141
carry-select adder, 108

carry-skip adder, 88
CDMA, 286
Chinese Remainder Theorem, 39, 213
  approximate, 194, 206
chips, 287
Code Division Multiple Access, 286
comparison, 194
complex-number arithmetic, 17
complex numbers, 40
compressor, 149
conditional-sum adder, 97
congruence, 6, 22
conventional multiplication, 138
conversion
  forward, 13, 49
    arbitrary moduli-sets, 58, 68
    extended special moduli-sets, 56
    modular exponentiation, 68
    special moduli-sets, 50
  mixed-radix, 213, 227
  reverse, 13, 213
convolution sum, 264
core function, 213, 234
Core Function, 197, 200
CRT, 39, 213
  approximate, 194, 206
cyclic property of residues, 65
DA, 264
delay element, 257
digital filter, 257
digital signal processing, 256

diminished-radix complement, 4
Discrete Fourier Transform, 272
distributed arithmetic, 264
division, 34, 201
   multiplicative, 201, 207
   non-restoring, 202
   residue, 13
   residue number systems, 34
   restoring, 202–203
   SRT, 202, 206, 221
   subtractive, 201–202
DSP, 256
dynamic range, 7
dynamic redundancy, 280
error correction, 17
error detection, 17, 279
error recovery, 279
error treatment, 279
extended special moduli-sets
   forward conversion, 56
fault prevention, 279
fault-tolerance, 17, 279
FDMA, 286
Fermat's Theorem, 32
finite impulse response filter, 257
FIR filter, 257
fixed-point number system, 1
floating-point number system, 1
forward conversion, 13, 49
   arbitrary moduli-sets, 58, 68
   extended special moduli-sets, 56
   modular exponentiation, 68
   special moduli-sets, 50
   special moduli-sets , 52
Fourier Series, 272
frequency-division Multiple Access, 286
harmonics, 273
high-radix multiplication, 142
high-radix reverse conversion, 248
IDFT, 274
IIR filter, 263
impulse response, 257
indices, 37
infinite impulse response filter, 257, 263

inner products, 264
interference, 286
inverse
   additive, 27
   multiplicative, 31
Inverse DFT, 274
MAC, 150
mixed-radix conversion, 213, 227
mixed-radix number system, 2, 13, 227
modular exponentiation, 68
modular multiplication, 137
modulus, 6
   redundant, 224
multi-operand modular adder, 63
multiplication
   conventional, 138
   high-radix, 142
   modular, 137
   residue, 12, 137
   residues, 29
multiplicative division, 201, 207
multiplicative inverse, 31
multiplier
   parallel-array, 145
   Wallace-tree, 146
multiplier recoding, 144
multiply-accumulate, 150
multiply-add, 150
narrowband, 286
negative numbers, 10
   representation, 2
Newton-Raphson procedure, 208
non-residue
   quadratic, 40
non-restoring division, 202
number representation
   residue number system, 24
number system
   fixed-point, 1
   floating-point, 1
   mixed-radix, 13, 227
   mixed-radix , 2
   positional, 2
   redundant signed-digit number, 5
   residue, 6

single-radix , 2
weighted, 2
one's complement, 2
  representation, 3
one-hot encoding, 14
overflow, 197
parallel-array multiplier, 145
parallel-prefix adder, 127
parity, 197
partial carry, 141
partial sum, 141
periodicity of residues, 66
period of residues, 66
pipelining, 263
polarity shift, 43
positional number system, 2
primitive root, 38
primitive roots, 37
projection, 284
quadratic non-residue, 40
quadratic residue, 40
radix, 6
radix complement, 4
radix point, 1
radix-r, 276
range
  dynamic, 7
recoding, 144
redundancy, 280
  dynamic, 280
  static, 280
redundant modulus, 224
redundant residue number system, 7, 42
redundant residue number systems, 281
redundant signed-digit number system, 5
redundant signed-digit set, 5
residue, 7
  cyclic property, 65
  period, 66
  periodicity, 66
  quadratic, 40
residue multiplication, 137
residue number system, 6

complex-number, 40
number representation, 24
redundant, 7, 42
standard, 7
residue number systems
  redundant, 281
residues
  addition, 29
  subtraction, 29
residue-set, 7
restoring division, 202–203
reverse conversion, 13, 213, 259
  high-radix, 248
  special moduli-sets, 237
ripple adder, 85
RRNS, 42
scaling, 36, 198
short period, 66
sign-and-magnitude, 2
  representation, 2
sign-determination, 197, 206
  approximate, 206
sign digit, 3–4
single-radix number system, 2
special moduli-sets
  conversion, 50
  reverse conversion, 237
spreading signal, 287
spread spectrum, 286
squaring, 150
SRT division, 202, 206, 221
stability, 264
standard residue number system, 7
static redundancy, 280
subfilter, 262
subtraction
  residue, 12
  residues, 29
subtractive division, 201–202
sum
  partial, 141
taps, 257
TDMA, 286
throughput, 263
time-division Multiple Access, 286
twiddle factor, 276

two's complement, 2
   representation, 3
unit circle, 264
unit delay, 257
VLSI, 258
Wallace-tree multiplier, 146
weight, 217
wideband, 286
zero
   fixed-point representation, 3–4
zeros of filter, 258